CANAL & INLAND
CRUISING

CRUISING

JOHN GAGG

Foulis

Haynes

For *Naec* (16ft), *Nike Two* (25ft), *Nike Three* (31ft), and *Nike Four* (45ft) − faithful companions, with their crews, over very many years. Forgive them for appearing rather often in the photographs, but they were always around.

A FOULIS Boating Book

First published 1989
© John Gagg 1989

Published by:
Haynes Publishing Group
Sparkford, Near Yeovil, Somerset
BA22 7JJ, England

Haynes Publications Inc.
861 Lawrence Drive, Newbury Park,
California 91320, USA

British Library Cataloguing in Publication Data
Gagg, J. C. (John Colton), *1916-*
    Canal & inland cruising.
    1. Great Britain. Inland waterways.
    Cruising − Manuals
    I. Title
    797.1

    ISBN 0 85429 698 0

Library of Congress Catalog Card Number
89-84301

Editor: Peter Johnson
Page layout and design: Chris Hull

Printed in England by:
J.H. Haynes & Co. Ltd.

Also by John Gagg
    Canals in Camera - 1
    Canals in Camera - 2
    5000 Miles, 3000 Locks
    The Canallers' Bedside Book
    A Canal & Waterways Armchair
    Book
    Observers Canals
    Looking at Inland Waterways
    (series)

CONTENTS

1 **"What do I do now?"**
A cautionary tale – You're not in a car – Spare time to find out
7

2 **What's a "Waterway"?**
Trying to discover – How did they all come about? Broad and narrow
11

3 **Boats and boats**
Shapes and sizes – Large and small – Trailing a boat – Let's look inside – Galley – Water supply – Saloon – Sleeping cabin(s) – Outside the boat – Gas
19

4 **All aboard!**
What have you brought? – Boarding the boat
29

5 **Those first few miles**
"Stoppages" – Other checks before setting off – Moving off – Steering – The "swivel effect" – Cross-winds – Reversing – Stopping – "Wash"
41

6 **Cruising along**
Safety, especially with children – Meeting other boats – Sound signals – Passing anglers – Rivers and tides – Going aground – Things round propellers – Turning round – Stopping and mooring – Knots – Throwing a rope – Switch off
47

7 **Bridges, tunnels and aqueducts**
Fixed bridges – Moveable bridges – Tunnels – Aqueducts
59

8 **Locks – at last!**
What are they? – Lock furniture – "Paddle-gear" – Opening the paddles
67

9 **Getting to work on a lock**
Locking "uphill" in a narrow lock – Rising in the lock – Downhill in a narrow lock – Broad locks – Guillotine gates – Staircases – Chief points to remember
79

10 **Looking around you**
Pubs – Lock- and bridge-houses – Maintenance yards – Maintenance boats – Commercial boats – Other boats – Numbers and notices – Stop-planks and stop-gates – Junctions – Looking at nature
91

11 **Some waterway highlights**
Some landmarks along waterways, in alphabetical order
109

12 **Yes, but which waterway?**
Potted descriptions of individual waterways
129

13 **Finding your way around**
Where am I? – Nicholson/Ordnance Survey Guides – Waterways World Guides – Pearsons Canal Companions – Other guides – Maps
153

14 **Bits and pieces**
Addresses – Museums – Books – Rallies – Hiring a boat – Buying a boat – Some waterway words
157

Index
167

You can cruise in the heart of the country . . .

6

1

"WHAT DO I DO NOW?"

A cautionary tale

Let me start with a cautionary tale – light-hearted perhaps, but it has its point.

It is to do with the heading of this chapter, which is a cry that I heard ringing across a canal long, long ago. It came from a hire-boat with a family aboard, and they had just arrived at the boatyard. There, still safely tied up, they had been in the process of learning something about the boat.

Unfortunately the "teacher" had been called away for a moment, and somehow they'd managed to start the engine, untie the boat, put it in gear, and start moving through the water. And that's when the cry rang out, with panic all round. For even at about 2mph, ten tons of steel boat on the move is a bit alarming if you haven't the faintest idea what to do with it.

Happy ending, of course. Someone managed to bellow across the canal and tell them how to take the boat out of gear again, and it then drifted gently into the far bank. There was hardly time, on a narrow canal, to shout out the technicalities of putting it into reverse, steering, and the other niceties necessary to get it under control and back where it came from.

Well, it won't happen to you, obviously – or will it? And one thing about this cautionary tale is a slight relief: whatever mess you make of things, it's highly unlikely – at that speed and on water – that you'll produce the results that a

You can cruise in the heart of Birmingham . . .

motorway crash can produce. All the same, ten tons is ten tons, and without some knowledge of how to handle it you can still bring about some unpleasant situations. It seems unbelievable that some of the small early hire-firms (now long gone) were alleged to say, "Can you drive a car? OK, then, off you go!"

You're not in a car

So off I go with the most important starter tip of all: driving a boat is NOT like driving a car. It's about as different as it can be in every respect except that you have an engine. And the biggest nuisance afloat is the chap who has driven like mad up the M6, jumped onto a boat, and set off with little knowledge of what he's doing. (The next biggest nuisance is the one who may well know what he's doing, but still expects to do vast distances every day, just like touring the Continent!).

Once you get car-driving out of your mind, though, inland cruising is a growingly easy thing. It's certainly a leisurely thing – just about the most leisurely way of passing the time ever invented, short of snoozing under a tree. But it's not as alarming a business as I may have made it sound above. In fact with a little patience anyone can learn how to set about it, handle the boat, look after it, work through the locks, and negotiate tunnels, bridges and aqueducts when met. There's absolutely nothing like it, even when pouring with rain. As for the M6, it might be in another world, with the Costa Brava and Majorca somewhere in hell.

Spare time to find out

The trouble is – as any hire firm will tell you – that most people hiring for the first time seem to have gone stone deaf. And however carefully they are told about what to do, it goes straight out of the other ear in their eagerness to be off. So something down on paper, such as a book, can at least be looked at later on.

There have been several books on how to cruise inland, and I remember well Michael Streat's excellent one in the days when he was

among the pioneers of hire-boating. Books appearing since then fall into two categories. They either seem afraid of driving people away, so tend to play down any snags or problems, including even the work necessary to look after a boat and navigate a waterway. The other kind seem to be designed as manuals for budding admirals/engineers, going into the most elaborate details that are really quite unnecessary. This type, too, seems to use a lot of sea-going terms ("port", "starboard", "bunkering", "helm", etc.), which purely inland boaters simply never use.

Children love waterway cruising!

I hope in this book to strike a happy medium. I'll try to take you in friendly fashion through every aspect of boating on our canals and rivers – but in as light and enjoyable a way as possible, without a lot of unnecessary technicalities. After all, the great aim is to enjoy cruising, not to make a miserable existence out of it (though you do see, now and again, some grim-faced chaps gripping tillers!).

One thing always stands out in cruising books, however: everyone has different ideas and tips. So I must stress that everything I say in this book is only MY way of doing things.

It comes from a lifetime of cruising all over the inland system, many times on some canals. But nobody differs so widely in their ideas as inland waterway enthusiasts. So if any expert disgrees with anything I say – fair enough.

The waterway system you can cruise in England and Wales is enjoyable and varied indeed. It runs from Godalming in the south as far north as Ripon. It stretches into Lincolnshire and East Anglia in the east to Llangollen in the west, with separate outposts near Abergavenny and even into Somerset and Devon. You can travel all over these waterways if you've got two

You can even try a horse-drawn trip. This is on the isolated Cromford Canal.

or three years to spare. And they're open to all, on either your own boat or on one of the large number of well-equipped hire-boats available in many places. But soak yourself well in it before trying it out – with hire brochures, books galore, *Waterways World* and *Canal & River-boat* magazines, and as a walker and gongoozler along the waterways themselves.

"Gongoozler"? is not in most dictionaries, but it means "an idle and inquisitive person who stands staring for prolonged periods at anything out of the common". The waterways are full of 'em.

So let's have a gongoozle first at these very waterways.

2

WHAT'S A "WATERWAY"?

Sorry, I keep on about "waterways", and "canals and rivers", but what am I talking about?

Maybe it's a good idea first to have a look at inland waterways before actually venturing on them. But if you're not much interested in how they got there, and how they differ from each other, skip this chapter!

Trying to discover

Sadly, you can't really learn much nowadays about inland waterways by going to the mighty boat show in London each January. Most boats there have little to do with cruising inland, and their sellers don't know much either about canals and rivers.

Many of the boat show boats are great

Mostly you are in peaceful countryside, and there's even peace in towns.

"gin-palace" yachts costing the earth, and of course are sea-going monsters. You have to search hard to find anything designed for cruising inland waterways. There are a few there if you look, though. The great differences are that they don't cost the earth (but can be quite luxurious, all the same), they don't have masts, they aren't usually more than 7ft wide because of canal locks, and they mustn't sit too low in the water or stick up more than about six feet above it. And to the artistic eye they are ugly and slab-sided, to lie nicely between the sides of a lock.

So don't take the boat show as much of a guide to inland waterways. No, take a car or bus and have a look at some. And find one of the many boatyards all over the place and have a look there, too.

What you're looking at is a remarkable higgledy-piggledy collection of canals and rivers, most of them linked up so that you can cruise over 2000 miles (and through 1400 locks) without leaving the water. Waterways lead you to the sea here and there where they join a river, but you don't normally head that way, since "inland" boats aren't suitable for waves and the sea. So you normally stick to canals and the various stretches of river which mingle with them inland here and there. The map on p.130 gives a rough idea of the extent of these waters, but more detailed maps and guides are listed on p.153 onwards.

How did they all come about?

Can you spare a moment for some history, for the background of these waterways is intriguing? I'm no great historian, but even the strongest history-hater has to be fascinated by the way in which this system of water-highways came about.

It was nothing to do with pleasure cruising, of course, but with the carrying of goods, and to a lesser extent, people. At one time any such transport could be a miserable business, with waggons, coaches and horses struggling or stuck in rutted "highways", especially in winter.

This is Worsley, from where the first true canal (after the Romans) was dug to Manchester, to carry the Duke of Bridgewater's coal.

Apart from main roads, most tracks were hopelessly maintained. But along rivers quite large ships and barges could travel with comparative ease.

This was all very well for those who lived by such rivers – apart from being flooded out at times. But what of the settlements far from a river, and only connected by bad roads? It was clear that sooner or later some bright spark would say "Why not dig artificial rivers?"

No new idea, of course, for "canals" had been dug from time immemorial, China onwards. It was really the Industrial Revolution that hurried things along in Britain, though. Coal and such raw materials as clay and ore tended annoyingly not always to be near a navigable river. Yet there was a desperate need to get raw materials to the factories, and goods away from them. Hence "canal mania".

How are you getting along? I hope this is relatively painless, for it seems a shame to talk about enjoying waterways without some small background talk as to why they are there.

So all over the country "developers" (as they'd be called these days) asked people to take shares in companies aiming to dig canals from A to B and C to D, usually to link with a river, but later to link canals already dug or being dug. Men galloped up from all directions carrying their bags of money, and there must have been few parts of the country, away from actual mountains, where some idea or other for a canal, crackpot or not, didn't emerge at some time or other.

So canals were dug – by rough chaps who came to be called "navigators" (hence the present term "navvies"). They frightened the lives out of people in remote villages as they passed that way, and the upheaval throughout the countryside must have been worse than when a new motorway smashes its way along nowadays. We can hardly imagine just what it must have been like to so many people whose lives had hardly been disturbed previously.

Far more than motorways, however, the canals settled themselves into their surroundings, but the great thing about them, from our present-day enjoyment point of view, was that each canal company had its own different ideas. They did in time get together to standardise (more or less) lock-sizes to allow boats to travel on different canals, but to this day the bridges, lock-houses, locks, lock-gear, warehouses (the few left), and other canal "furniture" differ intriguingly all over the country.

The company which dug the Macclesfield Canal, for example, preferred a pair of gates at each end of their locks, whereas most other companies with narrow locks had a pair at the top end, but a single double-width gate across the bottom end. The Trent & Mersey Canal Company provided interesting footbridges, rarely seen elsewhere, across the bottom end of many of the locks at the western end of the canal, with a split in the middle for the horse-towing-rope to pass through. The Calder & Hebble Navigation had its own unusual lock-gear, worked with a hefty club-like length of wood rather than a cranked handle.

And so on, and so on! It all means that we have inherited a wealth of differing canals, even though, nowadays, the companies have almost vanished, and the canals are mostly under the control of the British Waterways Board. The stamp of the many companies remains, so that cruising offers much variety. And unlike the railways, waterways have retained their original names, which all adds to the delight.

What we do need to be vigilant about, though, is the tendency to standardise things when they wear out. Lock-gear, especially, is a delightful study in variety. But there is a strong move to make it all the same from one end of the waterway system to the other. There's also a tendency to knock things down only too readily, so that old loading-bays, warehouses, lock cottages, etc., vanish without trace. Carting them off to the various excellent museums isn't quite the right answer.

Anyhow, much variety remains, and long may it do so. And as I have said, you can start

A lock on a "narrow" canal. Notice how the boat fits it sideways.

off at one end of the country and end up, even if years later, at the other end. But that's only for the retired or the prosperous. Most people will no doubt follow the pattern of so many friends that I know – they take the waterway system in pleasant and leisurely chunks. So either you'll hire a boat in a different area each year for a holiday or two, or get your own boat and moor it in a new place each year to explore a new area.

Broad and narrow

I've already referred to the different sizes of locks, and widths of waterways, and this will crop up all the time. So just a brief explanation of this.

Most rivers are quite wide and the locks built on them to make them more easily navigable were made big, too. The Trent, for example, now has locks 190ft long and 30ft wide, while the Thames has locks varying

A lock on a "broad" canal. The two narrow boats can come in and stay side by side.

16

between 109ft by 14ft and 174ft by 19ft 10ins lower down. So they can pass quite big boats.

When men began to dig canals, the great problem was always where to find the water for them. This had to be provided at their highest level, as it was always being used up through locks all the way down from there. So it mostly came from reservoirs specially built. Thus the canals in the higher areas were usually made narrower than rivers, and above all the locks

were made fairly small, too, so as not to need too much water.

Which brings us to our present-day "narrow canals". The locks on these are about 70ft long by 7ft wide (as you'll see when I come to the boats). Canals not so near the centre of the country needed fewer locks, and could perhaps be fed by water from rivers. So they often have wider locks. These are the "broad canals", and their locks are usually about 70ft long and 14ft

A ship lock on the R. Weaver, with a smaller lock alongside.

wide. Some rivers have locks about the same.

The bigger rivers, as I said, have wider locks than this, and the biggest locks of all are on the "ship canals" and the ship-carrying River Weaver. This last has locks up to 150ft by 35ft. The Manchester Ship Canal has some locks 600ft by 65ft. The Gloucester & Sharpness Ship Canal has an entrance lock 320ft by 55ft to the basin, but ships that size can't travel up the rest of the canal.

So "cruising on inland waterways" doesn't by any means imply that all these waterways are the same. They not only have locks of different sizes, and are themselves of different widths, but they have almost as many different characters as the old canal companies which had them dug.

3

BOATS AND BOATS

The beauty of inland waterways is that you can use all sorts of boats on them within certain limits, and by heavens people do! You see everything imaginable, from unofficial rafts on oil-drums to chromium-plated palaces (with microwaves, TV, stereo radio, bathroom, the lot); from canoes to paddle-driven live-aboards; from professionally-built beauties to home-made ones of varying finishes. And of course, especially on the wider canals, you still see commercial boats of different kinds and sizes, carrying goods more cheaply than any lorry or train.

There are some rules and regulations concerning safety and construction, but subject to these, anything goes, and private owners, anyhow, certainly let their hair down at times. The only other thing is that you must have a licence to use a boat. Most canals and some rivers are covered by one from the British Waterways Board, but there are also other areas such as the Thames, Broads, with a few more rivers and canals, coming under some other authority.

Shapes and sizes
The limits I mentioned above to boat-shape and size are those of the locks, depth of water, and height of bridges. The lock-sizes, of course, determine how wide and how long a boat may be, and most people cast their eyes on the locks

Canoeists are often afloat on canals, with a cheap licence available.

of the "narrow" canals. These are mostly in the central areas, and if you wish to cruise on these delightful waterways as well as on the bigger canals and rivers, the boat must fit these locks. That is, it must not exceed 7ft in width (usually built at 6ft 10in to allow for bulges), and 70ft in length. Boats of this kind (though often shorter than 70ft), are usually called "narrow boats", and real enthusiasts fall apart in horror if you call them "barges", which are 14ft wide or more. Aesthetically this narrow boat shape may seem at first glance ugly, looking like a long parallel-sided slab, compared with the flowing curves of sea-going boats. But it grows on you and it fits beautifully into a lock.

The snag is that 7ft of width, when used for pleasure-boat living, makes for strangely-shaped interior cabins, and if only the canal companies had settled on, say, 9ft wide locks, life would have been a lot easier for the designers of narrow boats when they began being used for pleasure purposes.

But that wasn't what they were designed for at first, and a long narrow container, as well as fitting snugly in the locks, could happily carry coal, grain, lime, stone, gravel and so on, without any bother about its shape. It wasn't until people started putting beds, tables, kitchens, washbasins etc., into narrow boats that they found things a bit of a squeeze, sideways, at least.

A rare sight — commercial narrow boats carrying goods.

Large and small

Before going on to these items, I must stress that, as I suggested at the beginning of this chapter, not all pleasure boats are traditional-shaped copies of the old narrow boats by any means. Indeed, when people first started to cruise for pleasure, the boats used were more likely to be of a sea-going shape, with curved sides, a deep keel, and hulls of wood. It did in time become clear that slab-sides were better for sitting alongside lock-walls, and bottoms were less likely to go aground if they were flatter and shallower. But boats were still often wooden, though soon glass-reinforced plastic, usually referred to as ''GRP'' or ''fibre-glass'', came into use, as it didn't damage or shrink as wood could do.

Then hire firms especially began to feel that steel was an even better material to stand up to the batterings of locks, bridges, and quaysides. But there are still very many GRP and even wooden hulls around, perhaps more often in private ownership. Smaller boats in particular

Small ''GRP'' cruisers, still common, but mostly privately owned.

are usually GRP, and a length of 25ft is very pleasant for a couple or a small family, for with a bit of give and take it can sleep four people. It is easily handled, and comparatively cheap, especially if you can find a well-cared-for second hand one.

The steel narrow boats are a good deal more expensive, of course, but commonly seen now both in hire fleets and with private owners. They can offer sleeping accommodation even up to ten or more people if they are the maximum length of 70ft. Many are shorter, and a 30ft or 40ft length can be quite roomy if used by only two or three people. A frequent dodge is for a couple to hire or own a ''four-berth'' boat. This will offer you the usual ability to turn its sitting-and-eating-area into bunks at night, so with only two people aboard this chore is avoided, as well as giving them more room to move about.

A scene on the Leeds & Liverpool Canal.
(Courtesy IML Waterway Cruising Ltd.)

A 25ft cruiser with outboard motor attached.

So there you are, preferably 7ft wide and from as short as 15ft to 70ft, all designed for people to live and sleep on, as well as cruise the waterways. If you stick only to the Thames, Great Ouse or one or two other rivers (and can afford it!), you can run a wider or even longer boat.

The bigger boats are usually driven by a diesel-powered engine, a bit noisy, perhaps, but diesel fuel is rather safer than petrol. Above all, for some odd reason, if it is bought by the waterway there is no tax on it, which makes it considerably cheaper than petrol. There's quite a bit of variation in prices, too, so ask around. Smaller boats are often propelled by an outboard engine which is attached to the stern. It can thus be unclamped and taken home if not too big. Outboards use petrol, or a petrol/oil mixture.

Trailing a boat

I must just mention here that one way of seeing a wider selection of waterways is to have a boat small enough to tow behind a car. In this way you can take it on a trailer to many different places, but of course you can't expect the interior space and comfort of a longer narrow boat.

There are many launching slipways all over the waterways, both private ones at boatyards and public ones which may have existed for some time. You need some advice on the right sort of trailer, and instruction on how to get a boat into the water at a slipway and especially on how to get it out again. Boatyards should be helpful at a price. If this appeals to you, you will undoubtedly be able to add a wide collection of waterways to your log book.

A small boat being launched at a slipway after being towed there by car.

Let's look inside

Now, what's inside these many kinds of boats? If you think of a house, then a boat of any size has pretty much the same things in it. But that's not as dull as it sounds, for they're all arranged in a different fashion from a house. And that's what intrigues so many people the first time they look over a boat. They see sinks and cookers, beds and tables, even radiators. And often you hear "Ooohs" and "Ahs" as they notice the ingenious ways in which civilised needs are adapted and fitted into these 7ft wide long shapes.

Grizzled canal users are sarcastic about some of this. When they hear of television sets, microwave cookers, telephones and central heating on boats they raise their hands in scorn. They then launch into stories of early hire-boats which provided spades to deal with the contents of bucket lavatories (or even boats with no lavatories at all), having to thaw out water-containers before use, walking across fields to get fresh water in such containers, cooking on paraffin stoves, and waking up in the morning soaked by leaks through wooden cabin-roofs.

Even the more tolerant users, indeed, may feel that portable telephones and microwaves go a bit far if a waterway cruise is to be anything of a change from the burdens of civilisation (which to my mind is its great beauty). Cruising can be a delight without some of our modern trappings, yet without exactly roughing it. But boat-builders and hire firms will say that some people demand more and more sophisticated equipment with their boats.

Good luck to them, then, if they'll only keep the stereo volume down! But the great majority don't go as far as all this. Most, however do these days expect a reasonable amount of comfort, so this is the sort of boat-fittings that I shall describe.

Galley

Well, this is a "boat-y" name anyhow, even if boats inland don't usually go in for sea-going words very much. But you can hardly call the cooking and washing-up area a "kitchen", as it's almost always part of the main sitting and eating area of a boat. Maybe there's a token low division wall between the two parts, but generally the cook can hand over the plates to those sitting hungrily waiting for what the good chef sends.

In a compact plan, there's the expected sink and draining board, perhaps, and a large or small working top. There's a cooker, ranging from two burners only to four with an oven underneath. There may be a fridge, often gas-operated, but sometimes electric (and thus battery-draining if you don't watch out). And there will be some sort of storage both for food and for crockery/cutlery (the latter being provided of course on hire boats). There'll be a rubbish-container of some kind, and there are regular places where these can be emptied.

I must say that I prefer the galley next to the steering position, so that the steerer can be handed hot or cold drinks easily! But sometimes it is at the front end of the boat.

The galley of a boat, with gas fridge, gas cooker, sink, drawers, etc.

24

Water supply

Just a word about the water that you need for the sink – and for the toilet area elsewhere. In the simplest boats you carry this about in containers, which you fill at the regular water-points along the route. Bigger boats have tanks under cockpits or beds, which you fill by hosepipe. Obviously even on the biggest boats such tanks can't be huge, so you shouldn't leave taps running as recklessly as at home. Many a family has done this, and run out of water the first day.

The water gets to the taps from these tanks by various electrical methods. Usually there's a pressure system, so that a pump builds up pressure which pushes the water through at the tap when you turn it on, the pump restoring the pressure after you turn the tap off.

Some of these systems tend to lose pressure gradually, and the pump automatically comes on for a few seconds to build it up again. That's all very well during the day, but can wake you up at night. When my own system developed a habit of doing this at one time, I put in a switch to turn off the pressure-pump if I wanted to.

Hot water? Ah, yes. Sometimes there's a gas-fired geyser as in houses. Or else you may have a calorifier just like the tank at home, which has its water-contents heated by the cooling-water of the engine running through it. This can be surprisingly effective, without the need for a gas-jet permanently lit inside the boat. But the water will have cooled by morning.

As I've already mentioned, many boats have central heating run from a gas boiler. Others may have it run by a heater using the same diesel fuel as the boat's engine.

Saloon

Next to the galley there is often the "saloon" which is the name given to the main living area. This is where there are seats of various designs, and I've even seen full-blown settees in use – though I feel they must collect a lot of damp in their springs and soft material. Most boats have foam rubber cushions, and the point is that almost always there is some system of turning all this into a bed or beds.

Ingeniously the table that is used with the seats during the day can be removed and used as a base for bed(s), with the cushions being jiggled around to make a mattress or two. The rest of the bedding – sleeping-bags, pillows, blankets or even duvets – is often stored in drawers under the daytime seats.

There may be a wardrobe in this part, for use by those who sleep there at night. And there may be drawers and cupboards according to the size of the boat. This, too, is the area where the TV, hi-fi, etc. live if you can't exist without them.

Evening's relaxation in the saloon. (Courtesy IML Waterway Cruising Ltd.)

Toilet area

Somewhere in the boat is a toilet section. This is often in the middle, so that those who sleep in the saloon and those who sleep in a separate sleeping cabin the other side of the toilet area can use it without disturbing each other. Bigger boats, with further sleeping cabins, may have a second toilet area.

Even baths have been seen on some boats, but a shower is more likely, draining into a tray under the floor to be pumped out electrically afterwards. There's a washbasin, of course, and drawers for storage.

Separately in this area is the lavatory. Once this was a mere bucket with chemicals, but the simplest these days is normally more hygienically designed and sealed in some way. It can then be transported for emptying at the flushing "sanitary stations" along the waterways. On the canals you need a British Waterways key for these places, which can be bought at boatyards. There are water-taps at these stations also, as well as at many other points, for filling your tanks.

Much more common in bigger boats is a "pump-out" lavatory. This flushes in some way, seemingly like a house lavatory, and a sealed "holding tank" in the bottom of the boat is pumped out hygienically at boatyards – at a fee.

Sleeping cabin(s)

And so to the rest of the interior. This is usually devoted to the sleeping cabin or cabins, and here is where you may start complaining about the old canal builders who decided on 7ft wide locks. For when the inside of a cabin is rather less than 7 feet wide, what do you do about beds (or bunks)?

Since there has to be room to move past (and dress and undress), then you haven't much space for king-size beds between the cabin sides. If you start thinking even of a 3-foot wide bed, you obviously aren't going to get another 3-foot wide bed on the opposite side of the cabin, or you'll have less than a foot between them to manoeuvre in. This in turn

Canalside "sanitary station" — with toilet, chemical toilet disposal room, and drinking water tap for hosepipes.

means that you would have to make the boat another six feet long to have another bed!

So, to keep the boat from extending by too many six-foot lengths, most sleeping cabins have quite narrow bunks across from each other, and it's still a squeeze in between them. Or, like railway sleeping-berths, they have upper and lower bunks. Indeed, some hire-boats that offer berths for 8 or 10 people do end up with sardine-like sleeping quarters. Conversely, comfort-loving private owners sometimes build themselves double beds, maybe across the cabin so that there is no through passage. Or more likely, the saloon seats are ingeniously arranged so that they extend to make a double bed at night, though without much floor space left.

Like the boats themselves, then, the sleeping arrangements produce all sorts of permutations and combinations. So if you are hiring a boat make sure you have enough room not only to sleep, but for all the sleepers to get undressed and to have somewhere to put their clothes. There should be wardrobes for this, and under-bunk drawers, but often they don't seem to have enough space. Which of course comes back to the business of not having vast quantities of luggage with you to start with.

Outside the boat

We've had a look at the innards of a boat, now let's go out. Outside there's the back deck, the front bit, the top and the sides, but some boats have a central open deck area instead of one at the back. Each of these areas serves a purpose.

At the back, or stern, traditionally-designed boats have a very small cockpit, with really only just room for the person steering. Mind you, this was next to the "boatman's cabin" with its stove, so the steerer could stand just inside the open doorway and keep warm. Most pleasure-boats now have a bigger stern deck so that the

Room to relax in the back cockpit. (Courtesy IML Waterway Cruising Ltd.)

steerer can have company and even a table can be set up when moored. Out there, anyhow, are the controls and the tiller or wheel. To obey canal by-laws, hanging over the stern there must be a fender, commonly of rope, to protect both lock-gates and your rudder from damage.

The front end of the boat must also wear a fender. At the front, too, there may be another cockpit offering a remarkably quiet sitting-out place where you can hardly hear the engine, and only the ripple of the water soothes you.

The roof of the boat is important. If it's made of steel an experienced boater can jump down on it from lock-sides, but often far too many people use it for sun-bathing as the boat moves along. Or youngsters fool about on it. Both activities can be highly dangerous, for many canal bridges only just clear boat roofs, and the point is that you never know when one of these is going to turn up round the next bend. Anyhow, swarms of people on a roof stop the steerer from seeing where he's going.

I like the minimum of anything at all on the roof, let alone people. So I don't have garden-boxes or mounds of logs or bags of coal there, or even decorative cans and buckets. They can all make life difficult, especially if you are taking ropes from one end to the other. So I stick to one short shaft (the name for a pole) and one long one, which we'll see in action later in this book when we go aground.

Fixed to the roof there should be handrails – perhaps set back a foot or so to grasp if walking along the side-decks. A golden rule is "one hand for the boat". That is, never let go of such a handrail as you walk along. And incidentally, never have the shafts lying against the rails, or they'll get in the way of your fingers trying to grip the rail.

Those side-decks: they're very narrow, and should only really be used by a confident boater to get from one end of the boat to the other, always with one hand gripping the handrail. Many a crew has slipped on such side-decks and been saved by his handrail-grip.

Besides the regulation front and back

Peaceful seating in the front cockpit on the R. Derwent. Notice the front fender.

fenders, most boats have fenders for the sides. These are not often of rope, but of white plastic. Their purpose is to protect the boat from rubbing or collision, but they are often over-used. Unless you have a fragile boat, or don't trust your steering much, there's really no need to cruise along with half-a-dozen fenders hanging over each side. In fact it's supposed to very unseamanlike to have fenders dangling as you are actually moving along.

It's quite common, though, to see GRP boats with their fenders down in locks. This is understandable, what with the rough lock-sides and perhaps the menace of a heavy steel boat alongside in a broad lock. But there is always the danger of fenders jamming the boat in a lock. Narrow locks may bulge a bit, and the extra width put on a boat by fenders may be just too much. Really, the main use of side fenders is when moored alongside concrete or stone quays. They stop a steel boat, especially, from keeping you awake at night as it grinds gently on the quayside. Fenders when not in use are best stowed away in a cockpit store, rather than left around on the roof or decks. There, they may be a hazard, or at best join all those others floating around the waterway system.

Gas

One other important item will be outside the boat – in the front or stern cockpit. This is the gas cylinder, in a compartment of its own, which will have a drainage-hole out of the side of the boat. This is in case there is any leak from the cylinder, since gas sinks. Needless to say, gas can be dangerous, and you must be sure the cylinder is connected correctly – and that gas-rings inside are turned off after use and when moving along. Make sure, if hiring a boat, that you are told all about the gas circuitry.

Many boats have gas-fridges, which of course should not be turned off. They should have double-safety fittings, which will turn off the gas if their light goes out, and which in any case vent out of the boat side. But this does mean that you can't turn off the gas-cylinder itself at night if you wish to keep the fridge in use. Thus take special care over turning off the other gas apparatus.

4

ALL ABOARD!

What have you brought?

So you're all set for a cruise, either on a hire boat or on your own boat. What I say will apply mostly to someone hiring a boat for the first time, but I hope some of it may be useful if you have bought a boat, too.

Now, is the car packed to the eaves with grips, cases, boxes, wet-suits, canoes, inflatable dinghies, fishing-tackle, etc? Well, perhaps you'll be hard pressed to store all that lot. A boat isn't as big as a holiday villa exactly, though most are much roomier than a caravan. Look at clothes, for example. There's no need to pack your entire wardrobe. The whole idea is to be as informal as you like. Thus it's much less hassle if you keep what you wear down to a minimum, though without roughing it. Take a decent outfit for the posh meal out, if you feel like it. But apart from that, casual clothes are the thing. Don't forget warm ones and anti-wet ones, though. "Shortie" gum-boots are helpful for wet towpaths and long grass if you're unlucky with the weather, but normally rubber or rope-soled shoes, not liable to slip on decks, are worn.

It may *rain, of course, so don't forget your wellies.*

Other personal items would be towel, soap and similar toilet things, and might include first-aid material, insect ointment, torch, cameras, etc., and maybe needle and thread. I'd also recommend an anti-sun barrier and perhaps lip-salve, for the air and wind can burn and dry up lips as much as the sun. I always take a book or two to catch up on bedtime reading (though often never open them after fresh air and exercise). And if you're wise you'll no doubt take books and games for children on wet days.

Apart from such things, most hire-boats provide everything else you need to live aboard. Bedding is usually supplied, though some people prefer to take their own sleeping-bags or even duvets. Life-jackets will be available if required, and are strongly recommended for non-swimming children especially. Quite a few boats are excellently appointed, even with TV if you must have it. The boat may carry bicycles, which are very useful for dealing with locks that are close together, and for shopping to a nearby village. But think where you're going to store them.

Talking of shopping, some firms will arrange for you to order food in advance to be waiting for you, though no doubt you'll also want to take your own favourite items of food and drink. In any case, you may find pleasure in local shopping – at hidden villages, for example, as I mentioned above. Don't always count on finding shops, though, for sometimes you seem to be miles from anywhere. Yet at other times the waterway actually passes shops and occasionally town centres.

There's a Sainsburys by the canal in Nottingham, for example, a huge supermarket by Rotherham lock, and in Lincoln, as I mention elsewhere, you can moor in a shopping centre almost between Woolworths and Marks & Spencers. But at other times you may have to walk or cycle a long way to a shop. So carry some reserves.

As to any possible "extras" for you to take – well, I've seen boats laden with canoes, or even towing some form of dinghy; and fishing-tackle shouldn't take up too much room (though make sure you are allowed to fish). It's up to you to take whatever you enjoy, subject to space. One word, though: as I've hinted earlier, the roof of a boat isn't really the place to carry things, for so many bridges may sweep them off.

Portable bike at work — for shopping or, in this case, for going ahead to get locks ready if others aren't about to use them.

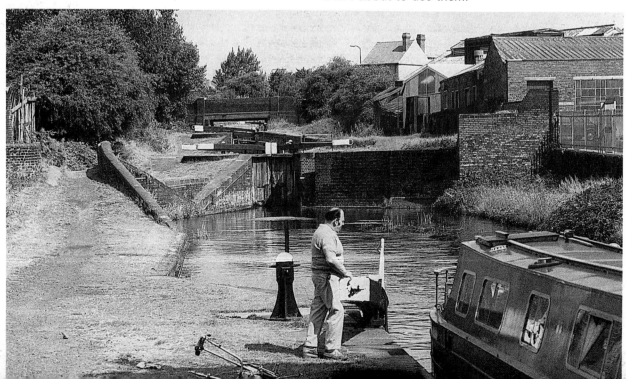

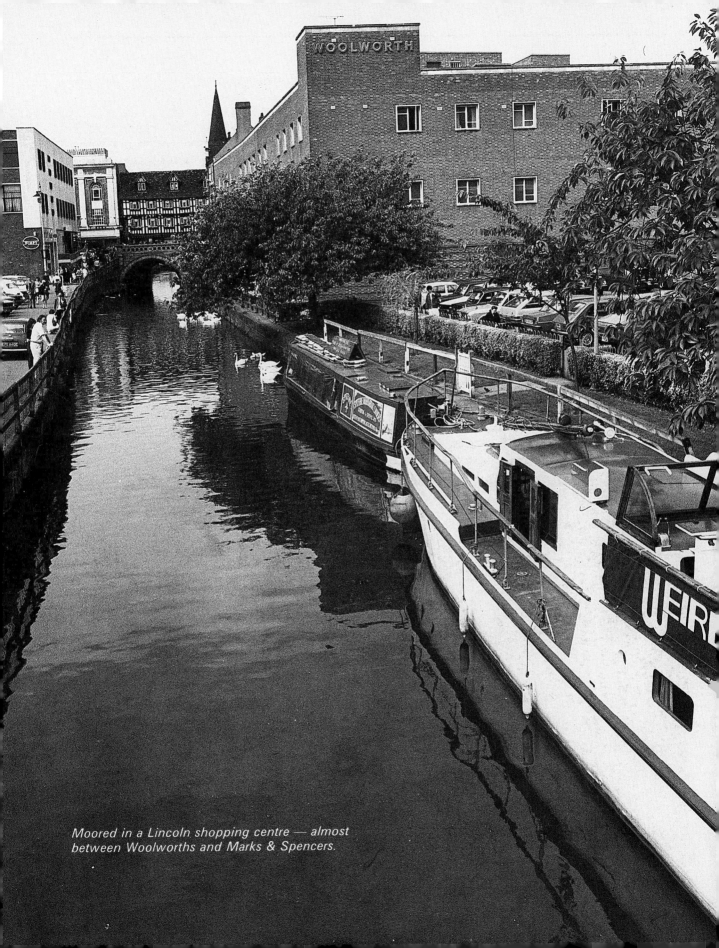

Moored in a Lincoln shopping centre — almost between Woolworths and Marks & Spencers.

Boarding the boat

So you arrive at the boatyard. It may be a harbour with piers and many boats, or it may be alongside a towpath. The boat could be far from your car, but usually it's near enough to carry things. So you do so, producing what looks like absolute chaos inside what was a tidy boat, as all your belongings are piled up. Don't despair. It's amazing how it can be tidied up, but not, I find, the first day!

Now don't nag the hire-firm to get you moving wihout being prepared to listen to them for a while. As I mentioned in Chapter 1, many firms complain that people can't be bothered to listen to hints and tips, as they're always so anxious to be off. Thus they end up making a mess of their first locks, and washing the banks away with their speed.

Seriously, do, if it's your first time, listen to the advice of the boatyard experts. They'll have sent you printed guidance which you'll no doubt have absorbed, and now they'll be ready actually to show you how to work the engine, the gas, the fridge, the cooker, the taps, etc.

They will show you how to start your engine, and how to put it into forward gear and change into reverse gear, as well as how to control the throttle. They'll explain quite a few other things, from daily checks to dealing with the gas and using the bilge-pump. They'll explain how you move away from the bank, steer the boat, and come in to land. Most will take you for a short cruise, and if possible show you how to work a lock. It's a sad fact, though, that this last important operation is often done very badly by people, and some of them swear that they weren't told what to do, even though they almost certainly were!

I hope that, besides the guidance of the boatyard, what I say later in this book will help, as, locks are perhaps the chief places where happy holidays go a bit wrong. Anyhow, whether you've absorbed all the advice or not,

Regular engine checks are necessary, and rather awkward too.

the time will come when you are turned loose in the great big waterway world on your own, with an expensive boat at your command, and 2000-odd miles of waterway ahead – if you've got a few years to spare, of course.

One of your crew pushes your bows (that's the front end) away from the side and gets aboard with the rope and mooring-pin (a metal stake which I'll mention again later). Someone else (or you) pushes out the stern not quite so much, so that you're pointing out at a slight angle, and also comes aboard with rope and pin, and you put the engine gently into forward gear. Opening the throttle a little, you're off! Unless, of course, someone's left a rope trailing in the water, in which case it may well get round the propeller and bring you to an ignominious stop. Ah well, let's get cruising.

Hotel boats pass through Manchester ▲

Off to the next lock on a useful crew member — a folding bicycle ▼

Main picture
Waiting for Tatenhill lock to fill — Trent & Mersey Canal

Inset
Peaceful winter cruising on the Northern Oxford

Below
Dwarfed in a Caledonian lock near Fort William

A blaze of gorse between tunnels on the Trent & Mersey

The Golden Steps to Heaven — Hatton flight in Warwickshire ▲

Lift-bridge on the New Junction in Yorkshire

Inset
Crossing Edstone aqueduct on the Stratford Canal

Main picture
Striking view down towards Wigan

Awesome sight beyond Ferrybridge lock ▲   *Inland meets ocean in Goole docks* ▼

5

THOSE FIRST FEW MILES

"Stoppages"

I mentioned "taking off" at the end of the previous chapter, but there's one small item before you do. Make sure that the route you've chosen is open!

This may sound silly, and normally (in summer at least) there's no problem. But things have to be done to canals just as they have to roads. And when a canal lock, especially, has to be repaired, the people responsible can't just close off half the carriageway and instal traffic-lights for one-way traffic. They simply have to close the canal. They call this, quaintly but accurately, a "stoppage". Normally in summer this doesn't happen, and such work is usually planned for the winter months (admire the maintenance crews!). But it's just possible that some emergency has occurred, so keep your eyes skinned for "stoppage notices" at junctions or other places well in advance.

Hire firms will tell you if there are any stoppages affecting you, but if you own a boat – at least for the canals which are run by British Waterways – it's safest to ring the special numbers they have for the latest urgent information. They are: (for the south and midlands) 01 723 8487; (for the north and midlands) 01 723 8486. If you cruise in winter, then you need the list of planned stoppages published by BWB, which you can get from where you get your licence. You can also find these lists in the waterway magazines.

Other checks before setting off

When you first start off, a hire-boat will have been checked for you. But I'll just mention here certain little checks to make each morning during a cruise (it's one way of getting out of the washing-up!) Not all need doing every day, and indeed some not at all in a week. But here are some reminders of items that you should at least keep an eye on. They will have been shown to you beforehand.

Certainly each day you should make sure the grease-feed to the stern gland is tightened, and check grease caps on the water-pump if any. Then you may like to clean the water-filter if you have one taking in canal water to the cooling-system, and dip your engine oil. And now and again, have a look at the acid-level in your batteries and top up with distilled water. If you have a closed-circuit cooling system, check on the water-level of its tank now and again also. And also make sure the bilges aren't full of water.

Moving off

Well, checks done and no stoppages, so let's get back to the "taking off" again, and really set out. I mentioned briefly this actual departure in the last chapter, and of course you don't want to make a mess of it in full view of a crowd, especially if it's your first time. So a reminder.

Start the engine and make sure no other boat is moving nearby. Then bows pushed off at an angle, front crew aboard with rope and mooring-pin, stern pushed off slightly and aboard with the other rope and mooring-pin, no rope trailing, into gentle forward gear, and away. This is usually straightforward, but if you've been moored in a very shallow area, make sure that the whole boat is well away from the bank before putting the engine in gear. The propeller will then of course start to turn, and if it's embedded in mud or worse it does it no good. But provided the boat is well afloat, the turning propeller starts to push it along. Panic all round now, sometimes!

Steering

The trouble is that with the most common canal boats you are standing at the very back, trying to see where the (distant) front is going. And it sometimes seems to have a will of its own.

This really is the single biggest surprise the first time you try to steer such a boat, especially if you are used to sitting in the front of a car to steer it. Standing at the back of thirty or forty or even seventy feet of boat seems a daft place to be, for the front seems to belong to another world. Keep the throttle down and see what happens. And if there are any other boats about, let them get out of your way first.

Normally you will be using a tiller to steer — an arm which is useful for leaning on when you get used to the whole business. The vital first thing to remember (and you may often forget this if you're a car driver), is that *in order to make the front end of the boat go to the left you must push the tiller over to the right, and vice versa.* And the second thing to learn is that the boat doesn't answer very quickly – unlike a car. And that's why there's often panic at first.

You've put the tiller *slightly* to the right, then, hoping the bows will turn slightly towards the left. They don't. Do not, I beg you, push the tiller even further over. Just hang on a moment – and sure enough, the bows will reluctantly begin to move over. When they do begin, straighten your tiller, or they'll move over too far.

That's really it, I suppose, *Tiller over slightly, wait a bit, see the bows begin to respond, tiller straight.* All very disconcerting for car drivers, what with appearing to have to move the tiller the "wrong" way, and the boat seeming not to take any notice at first. But there it is. That's probably why so many youngsters seem better at it than so many car-driving adults. Smaller boats may in fact have a wheel like a car, usually vertically-placed. In this case you use it not like a tiller, but actually like a car wheel. That is, to move to the left you turn the wheel to the left, and vice-versa. Many boats steered by a wheel have the steering-position somewhere near the middle of the boat and not at the back. On some rivers you can even hire boats which have the steering-position at the front, inside, where you sit down as if in a car. Personally I wouldn't be happy with this, since you don't easily know what's happening to the back end of the boat well behind you. And I'll mention in a moment the funny things that happen to the back ends of boats when you steer.

Practise steering, then, at slow speed, for some time if possible, till you really get the feel of it all. Since canal sides especially are often shallow, keep usually in the middle. In fact, you may find that after a time you can sense the "channel"; that is the deeper (usually central) part of the canal carved out by generations of boats, deeply laden at one time. Old hands reckon a boat will almost steer itself in this "channel", but I've not tried this.

You will of course move over (to the *right)* when meeting others. But if no-one's in sight, see how well you can hold an exactly straight course if you wish to, "sensing" the tiller as you move along. You find that you can just ease it one way or other almost instinctively after a time, to keep the boat straight. It helps, always, so keep your eyes on the very front of the boat, in line with something in the distance. In contrast, if you put the tiller over quite hard, the boat will, after the usual pause, respond quite sharply, so beware. And as soon as it starts to respond, as always then straighten your tiller, unless you wish to go round in circles.

Moving gently along with barely a ripple. Notice how far away the front seems till you get used to it.

The "Swivel Effect"

I must mention here yet another effect that makes steering a boat different from steering a car. It doesn't often matter very much, but it's an interesting thing, if a bit difficult to explain. Let me have a go, though.

When you turn a car wheel, the front wheels turn and the back wheels then follow them round, more or less, or cut inwards a bit. But when you steer a boat so that the bows go to the left, say, the stern in fact swings over to the right.

Let's just think about what's happening. The gadget that brings about the actual steering, of course, is not the tiller itself, but the *rudder* – a flat metal affair under water which is moved by the tiller. Normally it is sticking straight out behind you, and the water is flowing evenly past it on both its sides. But when you put the tiller to the right, for example, the rudder moves over to the left, and vice-versa. It is thus now facing the flow of water, which presses on it. This water-pressure on the rudder pushes it – and therefore the stern – over sideways.

So if you put the tiller to the right, the rudder moves to the left, and the water-pressure on it moves the stern over away from it, to the right. When the stern moves over, the whole boat then "swivels" so that the bows swing the other way, to the left! Clear as mud? If you look behind you, you may be able to tell that the stern is swinging over the opposite way to the bows, with the boat swivelling more or less on its middle.

As I said earlier, this doesn't usually matter much, but it may do if you are very near to another boat, the side of a bridge or lock, or a bank or quay. If you are near to something, and start to steer away from it, remember that when the bows try to turn away from whatever it is, your back end will swing *towards* it. So steer very gently away from any such items, or even push yourself off from it to be on the safe side.

This "swivel effect", by the way, is why you can't just steer away from the bank like a car when you start off. If you try that (and many do!) the back end just keeps swinging to the bank, hitting it, and thus bouncing the bows back in again. Hence the advice to push the bows till they face off at an angle first, and you can *then* drive straight off, at that angle.

Tiller over to the right a bit to start a slight turn to the left. Wait for the boat to answer.

Cross-winds

There's one other sometimes disconcerting point about controlling a boat: the influence of the wind. Cars are affected a little by cross winds, but they aren't actually pushed sideways on the ground. Even a mild cross-wind, however, will move a boat across the water sideways.

This again is something you can learn to sense, partly by the habit of lining up your bows with something ahead. If you find the boat seems to be drifting sideways, it will almost certainly be because of the wind. To counteract this, particularly with a strong wind, you need to steer slightly crabwise, pointing the bows towards the side the wind's coming from. Thus although you may seem to be steering towards the side of the canal, this is the only way to progress forward.

A really strong cross-wind can be quite alarming on a narrow canal, and it's only too easy to be blown on to a shallow side and be aground. For this, again see later. But even old hands prefer to moor in a gale.

Reversing

Now, one more tricky matter which is worth dealing with in the first few miles, reversing. This is most certainly very different from car-driving. In fact, there are many experienced boaters who avoid reversing if possible, since it's so difficult to do. And certainly some boats do it better than others because of the way they are built. So on the whole try not to have to go backwards if you can avoid it.

I've never been very clear about the mechanics myself, but roughly the turning of the propeller is always trying to move your stern over sideways anyhow, and it's worse when you are trying to go backwards. Also, the rudder combined with this is unpredictable, because of the water-flow. So you may well find it is impossible to steer a course if you have put your propeller in reverse.

The safest thing is to try to have your back-end pointing where you want to go before you actually go into reverse. Then, slowly, the boat may with a bit of luck keep that direction. If it veers off, then put yourself in forward gear for a moment, line up the boat again with a quick burst, and go into reverse once more, living in hopes. But I warn you, the steerer who can go far in reverse and still steer exactly where he wants to go is a very rare bird.

Stopping

It is in fact far more important to know how to STOP, for this is undoubtedly the single most vital difference from cars: **Boats have no brakes.**

The above-named reverse gear is in fact all that you have in order to stop, apart from hitting something solid head on. And unlike a car brake, you have to start using it as long as possible before you want it to work!

What is happening is that the propeller is turning and pushing water back, which tends to make the boat go forward. So if you turn the prop the other way, it pushes water forward and tries to make the boat go back. Since the boat may weigh ten tons or so, and water is a very fluid thing, nothing happens for quite a time as it ploughs happily on under its own momentum, ignoring the propeller trying to stop it doing so.

So use a good bit of throttle which should be *far more than in forward gear.* And slowly, slowly, the rapidly-turning, reversing propeller will slow the boat down. But I warn you that it takes some time. When you do actually stop the forward movement, be alert, as the boat will then of course want to start going backwards! So you have to judge just when to slow the propeller and then take it out of gear. It all sounds harassing, but it needn't be. But it is very important to commence the stopping procedure in very good time. You can't just "slam on the brakes".

"Wash"

One final point about your first few miles – and all the rest, for that matter. It is absolutely wrong to start opening up the throttle once you have the feel of steering. Not only is it against speed-limit bye-laws anyway, but (a) it's against all the pleasures of waterway enjoyment, and (b) it doesn't move you along much more quickly in any case, but merely uses up fuel, puts out smoke, and washes away the banks.

For some technical reason beyond me, however fast you turn the propeller, it won't push the boat along any more quickly *in shallow water.* It just hurls water back more quickly, throwing waves on the banks. On top of that, you'll spill the tea, beer, or chip-pan of every moored boat that you pass, not to mention banging them into each other if they are close together. You may see notices: "Speed Limit 4 mph". These are not immensely helpful, since no boat has any method of measuring its speed accurately anyway, unless you watch mileposts (if any) and use a stop-watch. But even if you could measure your speed, there are times when 4 mph is too fast and makes a wash, if the canal is shallower or narrower than usual. Looking behind you to see that there is only a ripple is a far better check.

WASH! The curse of the waterways, damaging banks and upsetting others. Look behind you often to make sure you're making none.

Sorry about that sermon, but the Golden Rule about cruising on shallow inland waters is to do it SLOWLY, without making a "wash" – those "waves" breaking on the banks behind you. It's bad for the banks, and bad for others. And apart from foolish behaviour at locks, it's about the only other thing that causes distress in an otherwise thoroughly relaxing recreation.

Gently past other boats, or they'll shunt into each other all the way along.

6

CRUISING ALONG

Here you are with the whole world of water-ways ahead (or anyway, a week or two of them to start with). I suppose I ought to plunge straight into locks, but there are several more things connected with actually moving along (and stopping for the night, shopping, etc.) that we might deal with first, things such as passing other boats, meeting anglers, going aground, getting things round your propeller, and so on. But if you are on a boat as you read this, and are likely to meet locks quite quickly, turn to Chapters 8 and 9 first.

Safety – especially with children

I mention safety items now and again, so may I touch on some here to start this chapter?

The foremost care for safety is if you have children aboard. If they can't swim they should wear the life-jackets which are normally provided by hire-firms. And you should make simple rules about not climbing on the roof or along the side-deck when moving along, and not hanging over the sides or putting hands over, especially at locks and bridges and near other boats.

At locks and moving bridges children will want to help, and should be encouraged to if well supervised. But there have been nasty accidents at both types of structure, trapping fingers and arms at bridges, catching fingers in lock-gear, leaving windlasses to fly off paddle-gear, and of course falling in. Any canal-item that moves should be carefully watched, such as the arms of lift and swing-bridges, the paddle-gear and the beams that open and close lock-gates. Without being too much of a wet blanket, I'm bound to say that it is possible for everything to be happy and relaxed at one moment, but for something unpleasant to happen quite quickly.

Much of the above applies to adults, too. Locks and bridges can play many tricks if you aren't alert. And the snag is that they are also places for happy conversation with other crews.

Life-jackets for children, on or off the boat. But keep an eye on them too.

But don't let this keep your eyes off the moving bridge or lock-gate, or the rising or falling boat in a lock.

And should children steer the boat? Older ones can soon get more skilled than you at it, but I've seen some shocking cases of tiny children at the tiller who can't even see where they're going, and of others left in unsure charge with adults inside. Surely younger children at least should only steer under careful supervision, for their sake as well as that of other boaters.

Meeting other boats

Now, even if you've managed to find a clear straight length of waterway to do your first simple steering, and your stopping and reversing practice, you'll meet another boat sooner or later. What do you do?

Two things, briefly. You slow down, and you move over to the *right*.

That is, don't think you're on a road, though the other chap may! There's no need to move too near to the bank, as this would bring you into the shallows. Just ease away from the middle, and aim to pass a yard or two from the other boat, no doubt exchanging a friendly greeting as you pass. (Yes, that's true – boaters are much happier people than car-drivers, though the exchange of greetings isn't quite as common as it once was – motorway influence, no doubt!). Having safely passed, you can move over into the channel again.

Meeting another boat. Slowly does it, and ease to the right, but not into the shallows.

Sound signals

Perhaps I ought to mention *sound signals* here, though I fear you won't find many inland boaters who understand them. They do appear in waterway bye-laws, however, and they are a means of trying to say things to other boaters, even if they don't understand you, which can't really be said clearly in any other way. Thus all boats must have some sort of whistle, hooter or horn, and though most people use them only at blind bends, etc., as if they are in a car, you really ought to know the "official" ways to use them. They're especially valuable if you're about to do something different from what the other chap is expecting.

The most likely action of this kind is if you intend to pass another boat on the "wrong" side ie., if for some reason you intend moving over to the left ("port") instead of to the right. The correct signal is two short blasts or whistles. But I fear the other person won't know this, so, undignified or not, stick your arm out as well!

Other signals are three short blasts if you're going into reverse, and four if you're going to turn round. For blind bends, a long blast every twenty seconds is the official signal, and a long blast should be sounded when approaching locks or bridges operated by keepers. (But for some reason, the lock and bridge signal on the River Weaver only is a long blast followed by a short blast). There are also rules for sound signals in fog, heavy rain and darkness, but maybe you'd better be moored then.

In fact waterway bye-laws make rather depressing reading in some ways, and might even scare you away. But many are technical ones, not perhaps applicable to you, even though you are officially supposed to know them. Indeed, some are regularly ignored, such as one that requires any vessel, when moored on a commercial waterway at night (Trent, Aire & Calder, etc.) to show a white light visible all round.

Passing anglers

Anyhow, back to ambling along: and during the fishing season you may come across anglers, especially at weekends when fishing matches may stretch for miles.

There has been a lot of controversy between boaters and anglers, but no purpose is served by enmity. Each to his own pleasure, and some give and take helps. It is obviously wrong, for example, for boats to roar past fishermen and upset their lines and keep-nets, just as it is obviously wrong for anglers to throw handfuls of maggots and swear-words at boats.

Unfortunately opinions differ about the best way to pass, except that everyone agrees that boats should pass very slowly. But I've heard some anglers object when I've moved over the other side as far from them as possible, saying that the fish are over there! Generally, though, I tend to keep away from them, though not right in the far shallows.

There are two particular hazards for boaters. Often anglers are hidden behind high growth on the towpath edge, especially in later summer, and you may not see them at all unless you spot the rod sticking out. The other recently-introduced danger is the long black fishing "poles" that are now common, stretching right across the canal. These have even electrocuted anglers by catching cables above them when lifted.

In theory the angler will move such a rod out of your way, either by pulling it back through the hedge behind him (with you praying that it doesn't catch in anything!), or by taking it to pieces as he pulls it back, or (most harassing) by lifting it up in the air at the last moment. This last leaves his bait wriggling just above your head. It is also risky if the hook catches on something under water and he fails to lift the rod after all. These new long poles are alarming things to see ahead of you, especially in a match when there may be dozens in a row.

Good luck with anglers, and don't forget to greet and thank them as you pass, even if you get no reply.

You'll often see rows of anglers. Pass them slowly, and keep over away from them if possible.

Rivers and tides

Now and again, if you are exploring the whole inland system, you'll come across rivers. They offer either cruising in themselves, or routes from one canal to another. In a few cases, such as the Trent, the Yorkshire Ouse, a short length of the Great Ouse, and in exceptional cruising the lower Severn and lower Thames, you will meet the tide below the lowest lock on the river. Rivers, especially their tidal parts, call for special care.

Even the mildest-seeming river has a flowing current, unlike most canals. Thus if your engine fails you will drift with the current,

and maybe over a weir at the next lock. So make particularly careful engine-checks before going on rivers. Have an anchor, too, taking advice on its size for your boat. Hire-firms will supply an anchor if needed, if they allow their boats on rivers. But make sure it is on the front of the boat, ready for use.

Those who aren't pretty confident swimmers are recommended to wear life-jackets, and especially don't let crew swarm about the outside of the boat. Allow for the current when slowing down, stopping and mooring. It's best to turn against the current to moor, and let it act as a brake. Tides can be the most disconcerting things. You not only have the river-current to think of, but maybe also the tide actually flowing the opposite way when it comes in – and pretty smartly back as it goes out.

Alarming experience the first time. An angler lifts up a huge black rod, always at the last minute!

On river navigations, have an anchor ready in the rare case of engine failure.

It's essential if travelling, say, on the Trent up to the Yorkshire canals, or up the Yorkshire Ouse to York, to take advice from lock-keepers about when to set out from and arrive at locks leading from the river concerned. For one thing, no-one in his senses travels against the tide, which can flow strongly. But if you travel down with it at the wrong time, you may end up stranded on a mudbank as the tide continues to go down. Again, arriving at a lock at the wrong time may mean that the water is too low for you to get into the lock. Or it may be flowing past so rapidly that it is difficult or even dangerous to try to get in.

So ask the keeper of the lock you depart from in the first place. Also, ring up (or ask him to ring up if he will) the lock for which you're heading, so that the keeper there will be on the lookout for you, and hopefully have his lock ready as you crab your way in.

The Trent especially has several tidal lengths used by inland boats. The mildest one is from Cromwell lock, below Newark, to Torksey lock to the Fossdyke. Then there's a stretch from Torksey past Gainsborough to West Stockwith lock from the Chesterfield Canal, and a length from there to Keadby lock for the Yorkshire canals. This last part is the trickiest, as the tide is more powerful lower down the river. Waterway guides will give you phone numbers for the locks, and you may even aspire to get hold of tide tables and work out for yourself the best times to set out and arrive. But be sure, as Summer Time and other variables can catch you.

Going aground

Back to more normal waterways, as I've said, I don't want all this to be a tale of woe, and I must always keep stressing how relaxing waterway cruising is. But things *may* crop up, so I mention them. One such is that you suddenly find yourself stuck on the mud, especially at the side of a narrow canal.

I've said several times that the sides of waterways may well be shallow, either from lack of dredging or because others travel too fast and wash soil in. So if you happen to wander too near to the side, especially away from the towpath, you may come to a gentle halt with your bows aground.

Don't burst into tears, for on a canal this isn't a terrible happening. The thing *not* to do is for crew members to rush up to the front to try and push you off with your long pole ("shaft"). Their weight will simply put the bows lower down still.

Instead try reversing the propeller if the stern itself is well away from the bank. This will often take you back off, particularly if everyone is at the stern to help tip up the bows a bit. Even rocking the tiller, or letting the crew rock the boat, may do the trick.

When you do come off, don't steer too sharply forward again, or the swing of the boat may bring the stern aground. Perhaps better to have someone go up to the bows when they are free, to push them off at an angle with the shaft before you set off again. But be careful with that shaft. Stay well in the cockpit when pushing with it, for it may slip. Alternatively, it may stick in the mud, so if the boat is moving away you may even have to let go of it.

If you can't get off with the engine, someone may have to go ashore and push you off with the shaft. But again, be careful where it is placed against the boat, or it may slip and go through a window. Failing all this, it may be a question of swimming across to the other side with your ropes knotted together, and pulling from there. Or you may find help from another boat, but be gentle with ropes and where you fasten them, and don't let the towing boat jerk too strongly. Some hire-firms, in fact, don't allow any towing.

Things round propellers

Another possible tale of woe is when you get something round the propeller. This may be polythene, a shopping bag or worse, or it may even be rope, barbed wire, a supermarket trolley, or bicycle wheel. All are not infrequent.

Usually it is polythene, and you can feel that you aren't going along as well as usual, and the engine seems to be labouring. It is especially difficult to slow down and stop, which can be risky. Something really bad will stop the engine. The first and often successful thing to do is to slow down, pause, and put the propeller into reverse (checking for other boats nearby). This quite often throws off any small item, though you may have to do it once or twice. If this doesn't work, then you'll have to stop the boat. Having tied up, remove the ignition key so that it is impossible for anyone accidentally to start the engine while you are at work on the propeller (it *has* happened).

Seek out the box-shaped hole called a "weed-hatch" (though weeds are rarely the problem nowadays). This is in the engine compartment above the propeller. You will have been shown how to remove its lid, but beware of damaging your hands on sharp metal edges, and then on the propeller itself. Roll up sleeves, gingerly reach down, and see what you've won.

There's no knowing what your propeller might catch! Here's a lovely piece of (now-oily) rope hacked off through the weed-hatch. THEN *put the hatch securely back!*

It may be possible to pull off rope or polythene, again careful of edges if it comes with a rush. Or you may have to get a knife to cut it, or even wire-cutters or a hacksaw for tougher items, in fact I ought to have mentioned these tools among things to bring. Needless to say, any such tools need careful handling when you are almost standing on your head in the engine compartment, with your arm under water.

Some things take a long time to remove. I remember a complete new tarpaulin one icy winter day (I thought my arm would drop off with frost-bite). But usually patience wins. If not, someone may even have to go in the water and work from behind the boat. At the very worst the boat will have to be lifted up at a boatyard, but this may never happen.

Once you have removed whatever you have won from the depths, don't throw it back in for someone else. Take it away with you to the nearest rubbish container. *Above all, put back the weed-hatch cover and secure it, or water will get into the boat when you start off again.*

Turning round
It sounds a bit early to be talking about turning round, but often you may feel like a short detour up a different canal, or travel up a branch canal, and need to turn the boat round long before the going-home turn. So let's look at it here.

Turning round can only be done, of course, where the waterway is quite a bit wider than your boat is long. This may be at a junction, or at various places along the canal if your boat is not a full-length one. And you will find "winding-holes" (pronounced as in the wind that blows), at places dug specially to allow full-size boats to be turned. Make sure the hole is a usable one, properly dredged, as some have been neglected.

Checking that there's no other boat too near, with your tiller hard over steer your bows into the "hole", slowing and stopping before they reach the end. Back off again (you may find the boat continues its swinging turn), and once more steer into the hole with the tiller hard over. By this time you should be well round, and another short reverse may have you pointing back where you came from. You may even be clever and use the wind to help, which gave the name to the holes and the boatmen's cunning.

With experience, if a canal is wide enough, you can turn your bows carefully into a soft bank, and go gently into forward gear with the tiller hard over. The boat will slowly swing until you can back off and set off the other way again. If in doubt, turn the boat by means of ropes ashore, someone pulling the bows along as the rest of the boat swings out and over.

Stopping and mooring

Starting off required certain tricks, to make sure you didn't bang back against the side. Stopping and mooring again is even trickier, and you see some fine old antics as people leap ashore, drop ropes and even get themselves wet, with angry captains shouting orders.

To avoid such a scene, first you make sure that the place you choose for tying up is a sensible one. You shouldn't moor near to a lock, a bend, or a bridge, at narrow points where others can't easily pass, or to private property. The towpath is all right otherwise, and you will find various official moorings also, and pubs with their own moorings for you. But try to check that the water is deep enough for you to come in.

Slowing down in good time is the second secret. Decide where you want to come in – either by a towpath where there seems enough depth, or at a quayside or pub mooring specially made for boats. Having slowed down, aim at a shallow angle towards the bank, and try to stop forward movement just before you get there. You do this, you'll remember, by going into reverse and giving the throttle a bit of a burst.

If you're lucky, the bows glide gently to the side, with the boat at a slight angle still (this angle is wise, in case the side is too shallow for

Coming in to moor (at The Gate!*). Crew member, coil of rope in hand and loose coils aboard to follow, will step ashore.*

your stern with its vulnerable propeller). A crew member steps ashore from the bows, carrying a mooring-pin and a coil or two of the front rope, making sure (a) that the other end of the rope is fixed to the boat, and (b) *that he leaves behind him some loose coils to follow him ashore* (or the rope will pull him back!)

Getting the stern in is the awkward part. It may be possible for you to be at such a small angle that someone can get ashore from the stern at the same time, but more likely you will move the tiller over towards the bank, and give a brief forward burst on the throttle, then into reverse again for a moment. This should have the effect of moving the stern in to the bank, then of stopping forward movement again. Someone then goes ashore from the stern also, again carrying mooring-pin and rope (with coils still aboard to follow him). Thus you have two crew members ashore, one at each end, holding the boat by its ropes – marvellous! Or, instead, somebody forgot something, such as a rope. (If you need to throw one to them, see later).

One other thing normally must be taken ashore, the hammer which is needed to knock in the mooring-pins. These "pins", or "stakes", by the way, are long metal spikes of various designs, to be knocked into the bank edge unless there are bollards or rings there already for you to tie up to. They must be very well bashed in, at an angle away from the boat, as passing boats draw you strongly and may pull them out. They must also NOT be put across the towpath, as this is dangerous to passers-by, and also against bye-laws.

The other points: the angle of your ropes, and the use of fenders.

I always find it best to take the front rope some distance forward from the boat at an angle of about 45 degrees, with the stern rope shorter almost level with its fastening. This all helps to stop the boat from sawing backwards and forwards as other boats pass. But to make this even less possible, some people put a third rope out at angle from the centre of the boat.

Trick for getting the whole boat to the side, especially if cruising alone. This is a centre-rope, which lies on the roof back to the cockpit when cruising. You can then stop, step off, and heave the whole boat in. Useful at times for temporary tie-up when locking, for example.

Now, if you are alongside concrete or some other hard edge, you may wish to hang fenders down to prevent noisy scraping of the boat in the night. Indeed, with a steel boat, that may be all you ever use fenders for. But don't forget to take them in when you leave, or you'll look un-seamanlike.

Knots

But what sort of knot have you tied? There are differences of opinion about what to use round the mooring-pins (or bollards, etc.). I always use two "half-hitches", making a clove-hitch, and it is impossible to describe in print how to do it. It's a sort of "flick-flick" movement of the hands, keeping the part from the boat pulled tight. I hope the photographs show this happening, but you will also find illustrations in most canal guide-books, too. I have always found this holds the boat well, and often you can just pull out your mooring-pin and the rope is free again when you want to leave.

Front rope well angled to help prevent movement when moored. Rear rope is at a shorter angle.
Sorry about the untidy rope.

One loop already pulled tight on the bollard, the
other being twisted ready (with the rope to the
boat tight).

Second loop on, pull tight. Add another loop if in
doubt.

Some enthusiasts scorn the clove-hitch, and go for a knot with various names, such as ''waterman's hitch''. This is made by taking a loop of rope round the pin or bollard, then passing it under itself and as a loop over the pin. Whichever you use, keep the part of the rope coming from the boat tight with one hand, or you will have tied up too loosely. Some boaters prefer to take the loose end of the rope back to the boat, tying it there too.

Throwing a rope

Sooner or later – either when mooring or certainly at locks, you'll have to throw a rope to someone, so let's just glance at this now. It isn't as easy as it sounds, and many a rope drops in the water. When it has been retrieved, it thus drenches everyone on the next throw.

Thinking back to the crew member stepping ashore when mooring, you may remember that I recommended him to take a coil or two of the ''shore'' end of the rope with him. He should then leave several coils lying on the deck behind him to follow him off (the far end of course being fastened to the boat). If he had taken the whole rope, the bit leading to the boat-fixing would simply pull him back aboard, or into the water.

The same principle applies in throwing a rope. Make sure one end is fixed to the boat, then take a few coils of the other end in your throwing hand. Leave the rest of the rope loosely coiled on the deck free from obstructions. Throw the coils from your hand, ''underhand'', and the rest should snake up after them, though this is more difficult if you are throwing upwards from down in a lock.

Throw near to the catcher, though not at his face. He should make sure he isn't hit, but be quick to catch the rope, or at least to put his foot on it where it falls. But again, you will see some fine old antics when ropes are thrown, not least when the other end wasn't fastened to the boat. But it's all part of the fun of cruising.

Switch off

So you're moored, probably without having to throw a rope, but maybe with a sigh of relief the first time. Check the ropes and stop the engine. This, with diesels, is probably done by pulling on a knob; I hope you've been shown. And by the way, don't forget to push the knob back again, or the engine may stay stopped the next time you try to start it (a common cause of mystery).

Take out the starter key and park it safely, and have a glance round to make sure all is tidy and safe. Now you can have your drink with a satisfied mind, sure of a day's work well done. And if the captain can also be brain-washed into doing the cooking, what more can anyone want?

Throwing a rope. Some coils in hand, others on the floor (or roof), loose and free to follow. Far end fastened to the boat. Throw underhand, but don't hit the receiver!

7

BRIDGES, TUNNELS AND AQUEDUCTS

Still cruising along, there are certain canal structures apart from locks which I'll deal with now, since a few tips on negotiating them are worth mentioning.

Fixed bridges

The bridges along canals can be a fascinating study, brick and stone, beautifully curved or ugly, often distinctive on different canals. Then there are the lovely curved "turnover" bridges which take the towpath (and once the horse) over the canal to the other side. Most bridges have numbers on them, sometimes varying on different canals, and some have names such as Milking Hill, Clogger, Adam and Eve. Indeed, bridge numbers and names are often the only way you can tell, with the aid of your guide book, exactly where you are, since canals are not full of signposts.

Canal builders seem to have enjoyed placing their bridges on bends so that you can't see if there's anyone coming the other way. Worse, they often made the ones on narrow-locked canals only wide enough for one boat at a time to get through. So if there's a bridge ahead, be wary. If you have a clear view, and no-one is in sight, carry on. If you can see someone coming, be polite. That is, if he is obviously nearer to the bridge than you are, wave him on and slow right down. Don't try racing. Holding back of course sometimes ends

up with an "After-you-Claude-after-you-Cecil" situation, but that's better than bumping into each other.

If you can't see beyond the bridge, slow down and sound your horn at some length (though people standing over diesel engines can't hear it). Then approach carefully, ready to stop and even reverse.

The actual passage of a "bridge-hole" on broad-locked canals is fairly easy, but on narrow canals, especially in a cross-wind, you're likely to scrape a side. It's intriguing to think out why. What happens is that you are instinctively allowing for the wind as you approach, but the bridge takes off any wind-pressure as the bows enter, and the boat veers over a bit. Then as the bows emerge, the wind catches them again, and they veer back. There's also the mysterious "paddle-wheel" action of the turning propeller, which is always trying to move the stern slightly to one side. This tends to twist the boat a bit as you pass through, even if there's no cross-wind. So don't be alarmed if there's a slight hiccup and you touch a side.

What's more important is to keep close to the towpath side under many bridges, because of a possible low slope of the bridge-arch on the opposite side. Here and there on the canal system are bridges where you can actually catch your roof on the side opposite the towpath, as many paint-marks bear witness. Do

◀ **Top:**
There are many gnarled old bridges about, often with character, but with slopes to be wary of.

◀ **Bottom:**
Towpath bridges can be delightful, as this pattern common in the Birmingham area shows.

keep your eyes skinned for cock-eyed bridges like this.

Approach as straight as possible, not too fast, but not too slow, or you can't steer. Glance over the side at the towpath edge, and try to move through just an inch or two from it all the way. That should automatically take care of the other side.

Movable bridges

On some canals there are low bridges which you have to move out of your way. The Oxford Canal has simple "tip-up" ones, and the Leeds & Liverpool has a large number to be swung sideways. The Llangollen has Dutch-like lift-bridges, with high top beams, and there are odd moving bridges on the Peak Forest, the Caldon, the Monmouthshire & Brecon, the Maccles-field, and a few elsewhere. On commercial canals in Yorkshire and Gloucestershire bridges are operated by bridge-keepers, either electrically or by hand.

To work your own, you have to draw in and put off a crew member. He will have to pull down a high beam for the lift-variety, and make sure it is kept firmly down (and thus the bridge firmly up) as you drive through. There will be very little clearance on the sloping side, so watch your roof and the wind. Swing-bridges are usually moved by pushing on an arm as on a lock-gate. Once through, draw in and pick up the crew, who will have lowered the bridge gently or swung it shut again.

One word of warning: never trust helpful children standing by such bridges. They may or may not know what they are doing. If they fail to keep a lift-bridge fully up as you pass by, it can damage your roof. Even a swing-bridge can catch your cockpit rail if it begins to close too soon as you pass. So if you meet these volunteers, at least make sure a crew member supervises them. Lift-bridges especially can tend to bounce back a bit after being first raised.

Easy through a "bridge-hole". Watch the wind, and aim to slip smoothly through, just missing the towpath edge.

Left:
Careful under a simple Oxford Canal lift-bridge. The bridge has been lifted by a crew member hidden by the foliage, and he is carefully sitting on the high lifting beam which he pulled down.

Below:
Easy through a commercial canal lift-bridge, raised for you by the press of a button, and almost always in good time for your arrival without slowing down.

Facing page, top:
Time and sheep wait for no man. This swing bridge will be opened by means of a long beam, when the last sheep permits.

Facing page, bottom:
Unusual lift-bridge in Huddersfield. This rises bodily when a large handle is turned.

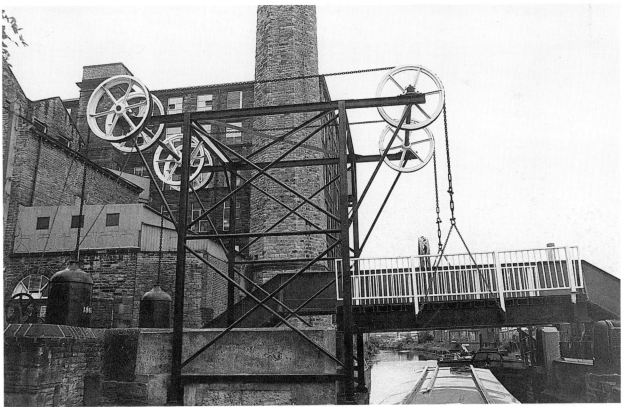

Tunnels

When a canal under construction came to hills, its engineer had to choose either to climb up them by means of locks, or go through by means of a tunnel, or a combination of both. There are 46 usable tunnels scattered throughout the canal system, though some are closed at times for repair. Some are quite long. The longest continuous tunnel in normal use is Blisworth on the Grand Union Canal near the Waterways Museum. It is 3056yds long. Another tunnel is a mere 25yds, shorter than some motorway bridges.

Inside, tunnels vary intriguingly. A few have been carved out of rock, or partly so, and have irregular sides. Others, especially two near Northwich on the Trent & Mersey Canal, have kinks in them. Tunnel insides may be streaked with chemicals or even the remains of smoke from steam days. Water may be running in from the roof or sides.

Passing through them can be an eerie experience, and some people frankly don't like it. But they can go in the boat, put the lights on, and think they're in a train (though trains don't take half-an-hour to pass a tunnel!). Most tunnels have no towpath: boats were "legged" through by men lying down and "walking" on the sides or roof. Often there is a footpath over the top, where the horses used to go; now it is useful for the nervous.

Apart from a few tunnels where there are clear notices, it is possible to pass another boat inside, even if this seems rather an alarming thought. You have a headlamp, of course, so make sure in time that it's working, and have someone ready to adjust it if it has been moved.

Short tunnel on the Chesterfield Canal; partly rock-sided.

Leaving the 3027yd Netherton Tunnel near Dudley, which has the luxury of two towpaths.

Don't let anyone travel on the roof. You might also, if steering, be glad of an umbrella for the hearty drips from some tunnels.

In the longest tunnels you may not even be able to see the far end at first, or it may be a tiny pin-point of light. It'll be there, though, so move along at a fair pace in the middle. It helps if your inside lights are on, so that you can gauge how far you are from each wall. When you see an oncoming headlight, try to estimate how far away the boat is, a very difficult thing, and of course boat-lights can't usually be dipped like car lights, and thus tend to dazzle. Slow down in time, and draw over to your right. You're almost bound to touch the wall a little now and

again, but you must keep moving gently in order to steer.

It will probably take longer than you thought for the other boat to arrive, and you won't know its length. It may even be a motor boat towing its "butty", i.e. two 70ft boats in line. So it's always a relief when you have passed, hopefully without a touch. Ghostly greetings in tunnels are intriguing, too. Don't, by the way, forget to switch off your headlamp after emerging.

Aqueducts

In striking contrast to tunnels are the occasional aqueducts which carry canals over a valley, road or railway. Some are hardly noticeable, unless you look down from an embankment and see a road disappearing underneath you and coming out the other side. Others are

remarkable structures. The most famous is Pontcysyllte in North Wales, 1007ft long on 18 pillars, and 121ft above the River Dee at one point.

The problem with such high, exposed aqueducts is that they have a very narrow channel, and the wind may twist the boat so that it is impossible to avoid touching the sides. Pontcysyllte, for example, has an iron trough made of 418 plates. This is nearly 12ft wide, but there is a towpath over some of this, so the boat hasn't much room. And of course two boats can't pass. Do your best, but don't be alarmed if you touch the side now and again. If you look down from some aqueducts away from the towpath, you can't even see any edge to the canal!

Stubby short aqueduct over the R. Idle in Retford.

Crossing the mighty Pontcysyllte Aqueduct in a 1007ft iron trough on 18 pillars, 120ft above the R. Dee at one point. You're bound to scrape the side, but don't worry.

8

LOCKS — AT LAST!

What are they?

I've been rumbling about locks here and there ever since the beginning of this book, and you may think I've been putting them off. In a way, yes. But I did think that the many things connected with actually moving along waterways might be better said first; though as I mentioned earlier, if you were likely to bump into a lock straight away, you could always turn to this chapter beforehand.

Anyhow, here we are at last. There's a curious contraption known as a lock, stuck across the canal or river in front of you, and an enjoyable bit of exercise ahead to get your boat to the other side of it. Let's just think about these ingenious items.

Locks, of course, are built to raise or lower the waterway level and boats on it to allow for slopes in the land. Rivers, having sources of water higher up, flow very gently on a downward slope all the time. But canals had to be dug on the level, or the water would all have run out of them, or at least, their feeder reservoirs built at the top level would have been drained. Thus when the land sloped, the canal had to be given a number of level lengths, or *pounds*, with "steps" between them.

If the land slope was great, there would be only short pounds and lots of steps. Indeed, at places the slope was so steep that the steps, the locks, had to run straight into each other with no pounds between. There are as many as eight such locks on the Caledonian Canal, for example, and five, four, three and two on quite a number of other canals. These are called *staircases*. Don't confuse them with places where there are strings of locks but with fairly short pounds in between. These are *flights*.

The locks you will find at intervals along rivers have a slightly different purpose. Most rivers naturally run on shallow beds, or with shallows here and there, and are thus not easily navigable by boats. To make them so, regular "dams" were put across them to build up water-depths above themselves. This, together with dredging, produced a navigable waterway. Originally these "dams" had ingenious ways of getting boats through them, often with a struggle against the current. But locks solved the problem and allowed boats easily to step down to the next stretch. The river-flow runs over weirs at the side of, or somewhere near, the locks. There are small weirs on some canals also, to cope with any surplus of water between locks, and to keep pounds between locks at steady levels.

One other point about locks, which I've mentioned earlier in connection with boat-widths. They vary in size, though fall roughly into two types. With some exceptions in Wales, the Middle Level Navigations, and a few elsewhere, they are either *narrow* (about 7ft

Staircase locks *run into each other, thus the gates between them are formidable, as here in Chester.*

A flight *of locks is a group close together but with short canal lengths between them. This striking lot are near Hatton in Warwickshire — the 21 "Golden Steps to Heaven". Notice the unusual paddle-gear.*

wide and about 70ft long), giving the name to *narrow canals*, or *broad* (about 14ft wide and about 70ft long, but wider and longer still on commercial canals and many rivers). Hence *broad canals.*

Most pleasure-boats are built to the 7ft width to allow use of all waterways. But those who stick to the Thames or some rivers have wider boats.

Lock furniture

Whether on rivers or canals, a lock is a sort of box of water, usually oblong, with various other necessary components.

There is the *chamber* itself, built of brick, stone, or even partly of wood in older ones. Then there are the solid doors, curiously called *"gates"*, across its ends, sometimes a single one, sometimes a pair. These can be swung open or in rare cases raised upwards, to allow boats in and out.

There will be a gate or a pair of gates at the "top" end, which is the end where the waterway outside is higher; and a pair of gates, or more rarely a single one, at the "bottom" end where the waterway beyond is lower. These gates close against a solid *sill*, or *cill*, to seal them at the bottom, and the sills at the top end can sometimes be a peril for boats, as I shall mention later. Most gates have *footplanks* fixed to them, for you to get across. Use them carefully, as they are often narrow and perhaps slippery. There is usually a handrail to use. Almost all gates have *balance-beams* sticking out over the land, to be pushed on in order to open and close the gates. On a few waterways both the gate-opening and the whole operation of the lock is electrified. You'll find this on the Thames and on commercial waterways in Yorkshire, for example, where you don't have to do any of the work at all. And on the River Nene, the Great Ouse, and one or two other places there are *guillotine gates*, which rise upwards as you wind a handle.

Then there may be *bollards* around to tie boats to; but beware if the boat is going down

Footplank across a lock on the Calder & Hebble. ▲
Notice also the "spike" to raise the paddle.

Facing page:
Guillotine gate on a R. Lark lock. This is raised a little to let the water out, then fully to let the boat out.

Balance-beam to open a lock-gate. Lean on the end gently, and the gate will move when it's ready.

in the lock! — and perhaps *ladders* in the lock-wall at some deep locks. There may also be *steps* or even a ladder by the side at the bottom end of the lock, to enable crew members ashore to walk or climb down to the lower level of waterway.

But before you can do any gate-swinging, the water-level has to be adjusted or you can't open the gates. And it is this adjustment that makes for the interest, enjoyment, work (and sometimes confusion) at locks.

"Paddle-gear"

Thus there's a vital further type of furniture at locks in the form of what's usually called *paddle-gear* (or sometimes *sluices* and, in the north, *cloughs,* pronounced "clows"). This is the actual machinery which enables you to use the lock, so look after it with care and guard it with your life. Its purpose is simple, even if its operation is sometimes complicated. It uncovers water-openings, either underground or on the gates, to let water in or out of the lock;

Ladders are often let into the sides of deep locks, for safety reasons. They are also useful for the steerer to climb up and help the crew!

just like that. The gears on the gates are called *gate paddles* and those in the ground are called *ground paddles.* Unless the lock is electrified as on big commercial canals and rivers, or a handle is fixed to the gear, you need a vital tool to operate this machinery. It is a lock key or lock

72

handle more commonly called a *windlass*, which is more precious than gold, and should never leave your loving care. I carry six, in fact, just to be sure, and true enthusiasts have brass ones, polished lovingly each day.

These bent crank-like handles are not as straightforward as you'd think. They have a socket-hole at the end, to fit on *spindles* on the gear, and there are in fact a number of different-sized spindles throughout the waterways. But usually two sizes of hole are enough to fit more or less all the spindles you'll find (though some may seem too loose or too tight). So, to avoid carrying two different windlases, ingenious people have invented some which have the two sizes of hole on the one handle.

◄

Hefty ground-paddle gear on the Caldon Canal. Notice the safety-catch gripping the cog. This stops the paddle from dropping down again when opened.

▼

Gate paddle raised near Wolverhampton. Again notice the small safety-catch at the landward end. The windlass will now, all the same, be taken off for safety.

I must say I prefer two different windlasses, and change them for different canals. Then I have a "Trent & Mersey" windlass to allow for the slightly different spindle-size there. And one day I'll treat myself to a "Kennet & Avon" sized one. To add to the variety, the Leeds Liverpool canal has such heavy paddle-gear in places that there's a special windlass available with a longer handle than most. This gives you extra leverage (and I use one elsewhere if gear is hard work). But be careful with one of these, since it may, where paddle-gear has been "modernised", catch your knuckles on balance-beams if you don't watch out.

So even a seemingly simple tool like a windlass can vary in all sorts of ways. And to help the variety, some people paint theirs red so that they can spot it if it is foolishly put down in the grass by mistake.

There are, just to keep you on your toes, at least two areas (other than those with electrified locks) where no windlass or turning handle is needed at all. At some locks on the Leeds & Liverpool Canal you grab a wooden arm and heave it up, thus moving a shutter out of the way under water. And the delightful Calder & Hebble Navigation in Yorkshire has much gear where you need to carry a hefty club-like length of wood, which is inserted in a socket to lever up the paddle. Useful in the old days if boatmen got into an argument.

So the variety of paddle-gear is amazing, even if there seem to be only slight differences between some types. The fact is, as I mentioned earlier in this book, that the original canal companies all had different ideas about almost everything, from bridges to lock-houses, and especially about paddle-gear. So the seemingly simple business of opening a hole under water has an intriguing variation of machinery to do it.

British waterways are, in fact, now standardising their spindles and sockets, so that's a help.

Raising a gate paddle with the famous Calder & Hebble handspike.

Opening the paddles

Basically, then, to open a paddle, the covering under water has to be pulled up (or in a few cases, swung sideways) to uncover a hole, whether in the gate or in the ground. Thus when you work the gear, you usually pull up a rod in some way which is attached to the shutter. Often there's a lot of cogs involved, with gearing of different types, pulling up a straight *rack* of teeth which pulls the rod up, which pulls the shutter up, which lets the water through.

In recent years a new type has been introduced which many enthusiasts dislike, partly because it doesn't seem any better than the old-established types, and partly because the variety of paddle-gear is one of the things which makes canals so fascinating. This new type works hydraulically (and is rudely referred to as "granny-gear"). You need to wind far more than with most of the old gear, and a tiny indicator rises in an enclosed casing to show you whether the paddle is "up" (open), or down. With the old unenclosed gear you can see from quite a distance whether the rack is up or down, and this helps to check both whether there's someone already at a lock as you approach, or whether you have (wrongly) left a paddle open after you have left.

Another objection to the new gear is that it is just as hard to wind it down as it was to wind it up. Whereas winding down the old gear is easy, since its own weight helps (but NEVER drop it!). But we're told that the new gear is safer, and certainly you can't catch your fingers in any cogs.

Whatever type of gear you come across, you almost always have to put the windlass on the spindle, and then use it as a handle to wind the gear open. You'll soon tell which way to turn. There's also a *safety-catch,* or *ratchet,* of

A rare collection of paddles for just one gate in Birmingham. Two ground and two gate, all raised and with catches on, and not lowered till the gate is opened. See the solid footplank, too.

some kind, which is supposed to stop the paddle-gear crashing down again after you have wound it up. You should usually put this safety-catch on before you wind, and get a satisfying clackety-clack as the gear goes open.

Never, however, trust this safety-catch. Many can slip off, so *always* take off your windlass after winding the gear open.

One other point about paddle-gear. In some areas the gear has to be locked in some way against vandalism. You will be told about this, and provided with a key, by hire firms.

It is a good thing that the "granny" gear has now ceased to be added, and that windlass-holes are being standardised.

Record-sized new-style "granny-gear" — hydraulically worked and not popular with enthusiasts. This is at a deep new lock in Bath. Notice the little indicator in the upright slot, to tell whether the paddle is up or not.

GETTING TO WORK ON A LOCK

Locking "uphill" in a narrow lock

Let's try a lock, and for the moment let's talk about the narrow locks around 7ft wide. And let's say that you're going "uphill" first; that is, you're moving from one level of water to a pound higher up. Much of what I say will also apply to locking downhill, and to using broad locks (over 7ft wide). But I'll mention the differences later.

As you approach the lock, check two things. First, is there anyone already in it? And second, is a boat approaching it from above? If there's a boat already in, then keep well back to allow its crew to empty the lock, and then to let the boat come out (the water coming out might wash you about if you stay too close). If there's a boat just approaching from above, it's entitled to use the lock first *if the water-level in the lock is already at the higher level.* If the water is at the lower level – i.e., yours – then it's your lock to use. Now and again you'll find others who, through ignorance or selfishness, don't observe these courtesies.

Say there's no boat near, or the water is already at your level. Then approach your lock, carefully come in to the bank, put a crew member ashore with a windlass, and wait for him to get the lock ready for you.

If the lock is "empty": that is at your level, he merely opens the gate or gates for you. He may be fooled into thinking he can do this when in fact the lock is a few inches short of being empty. But he won't be able to budge a gate! So in this case, as also if the lock is full, he has to use the paddle-gear.

First he checks that the gates at the other end are closed, and especially that the gear there is all wound right down (or he'll never empty the lock for you!). Then, putting his windlass on the spindle, and putting the safety catch on if possible, he winds away. Up goes the set of teeth (usually), or up goes the indicator in enclosed gear, and out comes the water from below the water-level. Then he opens the other paddle or paddles.

Watch your boat carefully to make sure it isn't washed around by this. If in doubt, keep it well back. Later, perhaps, you'll learn the trick of putting your bows actually touching the bottom gates of narrow locks before the paddles are opened, when the water coming out has the curious effect of keeping the boat there. But you must actually be touching the gates with your front fender to start with.

The gates can't be moved till the water-levels are equal, so, if you're the lock-worker, don't struggle. And don't lower all the paddles until after the gates are opened (leaks may foil you if you do). To open, lean gently on the far end of a balance-beam, and the gate will open slowly. You can't hurry it. *Make sure there is no-one – such as a kindly local helper – on the*

◀ *Opening a ground paddle.*

lock side of the balance-beam, or he'll be swept into the lock.

If there's only one lock-worker, he now has to go round the far end of the lock and walk back to open the other gate (there are usually two at the bottom end of narrow locks). Agile lock-workers leap across from one to the other to save the walk, but it isn't recommended.

Sorry it takes so long to write about what takes quite a short time in fact, but there are so many odd things to mention. Anyhow, the gates are open, and you can take the boat in.

You may be nervous about this at first, and rely on using your ropes. But with confidence you may well prefer to drive in slowly, going into reverse in good time to stop. It is in fact easier to drive into the bottom end of a lock than into the top end, since the lock *shoulders* channel you in. Gates to be closed now, and all paddles wound fully down after taking off safety-catches, and you're ready for the next step.

It's easier to drive into a lock than to heave the boat in with ropes — especially at the bottom end.

81

Rising in the lock

What happens now perhaps needs more care than any other aspect of using a lock. That is, the filling of the lock with water causing the boat to rise. The reason for this is that the water entering can wash the boat about in odd ways, and even pull it quickly to the top gates. It's far more relaxing when you are going down in a lock (unless someone has tied you up tightly!).

So beware. The steerer is sitting, of course, way down in the lock, and may not even be able to see the lock-worker. It's essential that the paddles at the top end are not opened too quickly, and without the chap on the boat knowing. In fact, it's often a good idea for him to throw a rope up to the lock side, and have it put on a bollard *back* from the boat, to prevent it being pulled forward by water-movement. Very long boats can put their bows gently up to the top gate and cill before a paddle is opened. I myself prefer to lie well back in a lock (with a 45ft boat), rather than put the bows on the gate at the front. But I make sure I'm not actually touching the back gates.

So the paddles are carefully raised, maybe not all at once. If the boat isn't tied, it is possible to use a touch of the engine in reverse to stop it moving forward. And if you, like me, are lying well back, make sure the boat doesn't drift forward. If it does, the forward pull of locks on some canals gets greater nearer the front, as the water rises. I've seen boats seemingly drifting gently forward, unnoticed, then rushing madly up as the lock gets fuller. So keep a wary eye open. One thing's for sure. If a boat does start rushing forward, holding a rope won't do much to stop it.

One other care. Some locks have odd protrusions or uneven stones in their walls, and it is just possible for a boat to catch on these. So both the lock-worker and the steerer should keep an eye on the boat all the time. If there is any snag, paddles should be wound down quickly to stop more water entering. Don't let me worry you, though. As with so many things, a gentle awareness is all that's needed.

One further word about the top paddles. On some locks there may be both gate paddles and ground paddles. If so, always open the *ground paddles first,* since it is possible that the gate paddles may be high enough to pour water on to the boat if it's near the top gate. When the water has risen somewhat, then try the gate paddles gently.

So, gradually, the lock fills and the boat rises. Sometimes it seems to take a long time for the last few inches until the water is level with that outside the top gate. Then again someone leans gently on the far end of the balance beam and the gate opens. Usually, by the way, it is a single big gate at the top end of narrow locks, though two on the Macclesfield Canal.

Drive out, then, and get ready to pick up the lock-worker(s). Their job is to close the gate, *always* (unless a boat is coming from the other way), wind down the paddles, and join you again, not forgetting the windlasses. Look back as you depart, in case a paddle has been left up.

A variation of this procedure is when there are locks close together in a flight, in which case a lock-worker can stroll up to the next lock and save you the manoeuvre of picking him up. Or, of course, he can use a portable bicycle.

"Downhill" in a narrow lock

Most of the actions I've mentioned above relate to locking downhill also, as well as to broad locks. The paddle-opening and closing, the gate-opening and closing, and the careful watching of the boat, for example, are general to all locks, whether going up or down. But now I'll mention differences in locking down in a narrow lock.

It's rather more important to check that there is no other boat lurking below the bottom gates, for example, either ready to come in because it's "their" lock, or, after you are in and ready to empty the lock, liable to be dashed about by your water. It's also not so easy to drive into the top end of a lock, with no handy shoulders to guide you, as the edges are not so

Drive in at the top end if you can, but it's not as easy as at the bottom.

easily seen as the high shoulders at the bottom end. Nor can you put the bows to the gate beforehand, as it has to open towards you. You can drive in, though, with care, if the wind isn't too strong to twist you.

When you're in, with the gate closed and the paddles wound down, beware of one thing before the bottom-end paddles are raised. *Don't stay too far back in the lock (too near the gate behind you)*. The (invisible) cill under the water will extend some way forward, and the boat could settle down on it if it is too close to the gate. So keep an eye on this or your propeller and rudder could be damaged.

Otherwise, a boat settling down in a lock doesn't move about as it may do when rising. You do need to watch for any unevenness in the lock walls, and the boat will perhaps drift gently forwards, but of course it mustn't be tied tightly to bollards or it will hang up!

When the water is level, open the gates as usual, and then lower the paddles. It's easier to sit in the lock-shoulder just clear of the gates to pick up your crew, rather than having to come in to the bank. So you're off.

Beware the cill behind you when going down in a lock (this is a broad lock cill). ▶

Opening a gate-paddle at the bottom end of a narrow lock.

Lean on the end of the balance-beam to open a gate. This is a single gate at the bottom of a BCN lock. Most narrow locks have two gates at the bottom end.

Broad locks

Now for differences in broad locks, first of all dealing with the 14ft wide ones common to many canals. The chief fact is that a narrow boat doesn't fit into them like a hand in a glove, as it does in narrow locks. Indeed, the best thing to aim for is to share a broad lock with another narrow boat about the same length as yours. Thus you can sit side by side, the steerers chatting to each other as the crew does the lock-work. Or so wives say.

Gates are heavier at broad locks, and there will be a pair at each end. If there's no boat to share with, there's no need for the lock-worker to open both gates (it's a long way round for him or her), since a narrow boat can get in and out through a single gate. A single narrow boat in a broad lock should lie against one wall, especially when locking uphill. In fact it's a good idea to use ropes on bollards to keep it against the wall, for the incoming water can dash a boat across the lock. A tip, when locking up on most broad canals, is to open the ground paddle first on the same side as the boat is lying. As often as not, the water coming in then holds the boat to the wall.

On the Thames, Trent, Severn, some other rivers and canals, such as the Caledonian and the commercial canals of Yorkshire, locks are bigger than the 14ft by 70ft of ordinary broad canals. Some are very much bigger. Most of these are controlled electrically by lock-keepers, sometimes from high cabins. The biggest locks hold a number of boats of different shapes and sizes, and Thames keepers especially are skilled at packing them in. This is where you use fenders and if necessary ropes.

I find the huge locks in Yorkshire surprisingly gentle, and can even just stay in the middle without being washed about. But do as the lock-keeper says! Commercial locks usually have traffic lights to tell you when to enter. These must of course be obeyed, for you will as like as not find a vast barge emerging. Mind you, I enjoy the pleasure of sharing a waterway with commercial traffic, for their captains are considerate. Even among the ships on the Weaver, Trent and Gloucester-Sharpness Ship Canal, life seems quite relaxing. Maybe this is the contrast with motorways, for I'd rather meet a ship than a juggernaut any day.

There's plenty of room in a Trent lock, if necessary, for quite a number of canal boats. ▶

Facing page, bottom:
You'll find traffic lights at commercial locks, as here on the Sheffield & S. Yorkshire Navigation.

Life is much easier in broad locks if two narrow boats lie side by side. ▼

Guillotine gates

As I mentioned earlier, on the River Nene, the Great Ouse and one or two other odd places, locks have a huge steel gate at one end which rises in the air. This acts as its own paddle, since the moment it begins to rise, the water runs out under it.

There is usually a handle already there to open these gates, and the Nene gates have locking devices for which you need a key (obtained with your licence). You raise the guillotine only a few inches at first, then let the water flow out until the lock is level with the water outside. Then you can raise the gate the whole way, a hard job, but it has to be done or the boat can't get underneath. And it's just as hard winding the gate down again; one Nene lock needs 150-odd turns each way.

On the Nene especially there are rules about how the locks should be left: guillotine gate up, which means that each time you use a lock you have eventually wound the gate both ways, whether going up in the lock or down.

Winding a large handle to raise a R. Nene guillotine.

The captain climbing up the lockside ladder — with rope and windlass — to help with a Nene guillotine and the top gates.

Staircases

There remains the use of staircase locks. These, you'll recall, are where one lock leads straight into the next – perhaps up to five at once (eight on the mighty Caledonian in Scotland). There are quite a few staircases on the Leeds & Liverpool.

There are two main points to remember. First, on narrow canals you can't enter a staircase if a boat is already coming the other way, since you can't pass each other. (You *can* do this with narrow boats in broad staircases, but few people seem to fathom how!). Second, you can't empty a lock usually unless the lock below is empty to receive the water. There are exceptions to this with side-ponds and over-flows, but it's a good rule to stick to.

Thus, to take the last point, if you are about to do down a staircase of five locks, you must have the lower four empty before you can empty the top (full) one which you have entered. Then you use the water from that lock to take you all the way down. Going up, the opposite (more or less) must happen. That is, when you are in one (empty) lock, the one above must be full in order to fill the one you are in and so on upwards. All this can be confusing, but fortunately there are usually lock-keepers around at the busiest staircases such as the famous Bingley Five in Yorkshire, and Foxton and Watford on the Leicester line of the Grand Union Canal.

Emerging from the bottom of the five Bingley staircase locks.

Chief points to remember

I think that's about it over locking – more than enough, you might say. But it is the single most important thing to get right on waterways – especially canals – and it's possible to make a complete mess of it, to your sorrow and that of others. Yet many people come to think that lock-working is the most enjoyable part of a cruise.

Could I end this chapter, then, by listing what I feel are the most important items to remember?

1. *Never, never hurry at locks.*
2. *If the lock is "ready" for an oncoming boat, allow it to use it first.*
3. *Always check that the paddles at the opposite end are fully down before opening those at your end.*
4. *Watch the boat carefully at all times, including when it is outside the bottom of a lock as water emerges.*
5. *Especially control the boat when it is rising in a lock – perhaps keeping it towards the back end unless you have a very long boat.*
6. *Even when going down in a lock, watch the boat, as it may catch on protrusions. Especially, don't tie it up!*
7. *Beware of "helpers" and spectators at locks. Make sure they are in a safe place, especially near swinging balance-beams.*
8. *Keep control of children (and dogs!).*
9. *Always put safety catches on paddles when wound up – or before you wind if possible – and take off windlasses.*
10. *Never let paddles drop down, or they may be damaged.*
11. *When going down in a lock, don't have the boat too far back, for there is a cill under water by the gates.*
12. *In broad locks, try to share side-by-side with another boat. If alone, keep to one side and use ropes.*
13. *Again if alone and rising in a broad lock, try opening first the ground paddle at the same side as your boat.*

My secret weapon to get from the boat in a deep narrow lock to help with the work. It lodges by the boat handrail, but you've got to be quick climbing up!

14. *Alone at a broad lock, save work by using only one gate to go in and out.*
15. *Remember at staircases that you can't empty a full lock into another full lock.*
16. *Obey traffic-lights and lock-keepers where you find them, especially at big commercial locks.*
17. *Close all gates and paddles before leaving, and look back to check. And have you still got your windlass(es), not to mention all your passengers?*

10

LOOKING AROUND YOU

If you're a new steerer you'll probably find that for the first day or two of a cruise your eyes are glued most of the time to the bows and the waterway in front. Mind you, on some narrow canals it might be a bit risky for your eyes to wander too much. But once you begin to relax you realise just what a lot of things there are around that you can never notice driving on roads. Indeed, many of them aren't to be seen along roads anyway.

In this section, then, I'll mention some of the general things to be seen at 3mph, or when moored, those things especially enjoyable along waterways. So keep your eyes skinned. In the next chapter I'll also look at some particular landmarks along waterways.

I've mentioned some of the canal sights already, of course, in connection with the actual cruising, tunnels, locks, and so on, and I made a number of comments about such structures in the cruising details earlier. Among the facts worth studying about tunnels, for example, are their locations, how long they are, their special history if any (e.g. Harecastle's fascinating story), and so on. Even if tunnels don't grip you, you may become quite hooked on locks and their gearing. And in fact the variety of different gear, gates, balance-beams, ladders, bollards, footplanks and so on is quite amazing. Many years ago I started to photograph this variety, and I haven't finished yet. But the powers-that-

be are getting rid of some of the most intriguing items, so pester them to stop.

Again, aqueducts and bridges, especially the latter, can be quite a pleasure. The bridges, as I said when dealing with steering through them, vary with different canal companies and some are artistically delightful, while others are ugly – notably the latest replacements. So I won't say more about tunnels, aqueducts, locks and bridges here, though special examples will turn up in the chapter on Waterway Highlights. So let's call at a pub instead.

Pubs

The old boatmen got thirsty working locks and maybe legging through tunnels. And perhaps they had to stop for the night where there was provision for their horses. Hence waterside inns, often with stabling attached.

Most of the remote ones have gone now, but many which were also alongside roads remain. Large numbers of these adapted themselves to the road traffic, and in fact often their canalside ground, until quite recently, was used as a dump for empties and miscellaneous rubbish.

It's interesting, then, to see the gradual changeover in the last twenty years or so as waterways have grown in popularity. These pubs have turned round again, and although

The well-known Swan *at Fradley Junction, where the Coventry Canal joins the Trent & Mersey.*

they don't exactly ignore the car trade, since they have to live in winter, they have certainly set out to attract the boat and towpath walker trade as well. Many have installed mooring-lengths, with rings or bollards to tie to (though there often seem to be boats permanently moored there!).

Quite a few have even changed their names to suit. the *Globe* near Weedon on the Grand Union has become the *Narrow Boat*, for example (but has a Chinese restaurant!). A few other *Narrow Boats* have blossomed up and down the waterways, complete with pictures-que (or, sometimes, all wrong) inn-sign.

Many, though, have always had waterway names, such as *Wharf, Boat, Barge, Swan, Bridge, Lock* (and *Big Lock, Kings Lock, Three Locks,* etc.), *Navigation, Canal Tavern, Jolly Boatman.* Some have more sea-going names – *Ship, Anchor, Pilot, Jolly Tar* and *Jolly Sailor,* and even a *Rock of Gilbraltar* and a *Cape of Good Hope.*

I once started the long pointless job of counting the various names, but ran out of patience when I'd got to 26 *Navigations* from Gnosall on the Shropshire Union to Blackburn on the Leeds & Liverpool, 19 *Bridges* from Llanfoist on the Monmouthshire & Brecon to Etruria in Stoke-on-Trent and 18 *Boats* from Berkhamsted on the Grand Union to Hayton on the Chesterfield. Among the odd names there's

You'll find more and more pub-signs with canal flavours.

a *Plum Pudding* near Rugeley, a *Pyewipe* (old word for peewit) on the way to Lincoln, a *Friendship* and a *Unicorn* on the Trent, a *Hop Pole* on the Chesterfield and a *Malt Shovel* at Shardlow on the Trent & Mersey.

All right, I'll stop, for of course it's what's inside that is more likely to interest you. And I'm bound to report that although some have tried to make a genuine canal atmosphere with pictures and even bits and pieces of canal furniture (and have provided facilities for children), many have just adopted the loud colours and loud musak atmosphere that has nothing to do with canal peace. Perhaps someone should try to compile a "Good Canal Pub Guide" which deals with atmosphere as well as beer.

All the pubs alongside your route are listed in the main guides, so that gives you a lead, but don't forget those a little way up the road in some remote village. They may be worth seeking out.

Lock- and bridge-houses

As I've mentioned often, different canal companies built their canals differently. Besides the varying locks, etc., another noticeable variety lies in the lock-keepers' houses. There were of course far more keepers than now, and many a lock had its little cottage alongside. Some still do, especially where there is a whole group of locks, and the cottages differ greatly. I think of a little squat one by the two locks at Brighouse on the Calder & Hebble, and another by the three locks at Salterhebble on the same navigation. Some of the Shropshire Union ones, too, seem to squat there by the lock.

Perhaps the most striking lock-houses are on the southern Stratford canal, with their distinctive barrel-shaped roofs, and there's a doubtful story that they were built that shape because the canal builders knew only how to build bridges and tunnels. Most of these are privately owned now, some lovingly looked after. Curiously, the lock-houses suddenly change as you go down this canal, and become more upright, ordinary-shaped.

The Thames, as you would expect, has much more superior looking houses, and there is a long tradition of keeping their whole surroundings beautiful with flowers. But on the canals some of the remaining lock-keepers enter the annual competition for the best-kept lock, and you see them mowing and gardening in their spare time. Sawley locks on the Trent look good, for example, and the whole flight at Adderley on the Shropshire Union has figured in the prize-list.

The BCN had a large number of buildings where its employees lived, and not only at locks. They too are often squat, and they carry large numbers on the wall.

There are some solid-looking houses on the eastern Trent & Mersey, and the Grand Union ones also look rather undistinguished. But matching the Stratford ones in unusualness are some houses that aren't connected with locks at all, but with bridges. These are the bridge-keepers' houses on the Gloucester & Sharpness Ship Canal, which look as if they've come from some ancient civilisation, with pillars in front.

A lock-keeper's striking little house at Salterhebble on the Calder & Hebble.

Of interest, too, are the lock-cabins on big canals and rivers, from which the keepers control the locks electrically. From high up, to give them a view, the men watch the passing scene. You'll find these on the Trent, Severn, and the Yorkshire canals, with some rather pleasant new ones on the Sheffield & South Yorkshire.

As I said, many lock-houses have gone, where they were far from the usual services. Others house waterway employees who do not deal with the locks. Yet others, as on the Stratford, are now occupied privately, often by canal enthusiasts, who may well set up a small shop for passing boaters.

One of the several bridge-keepers' houses on the Gloucester & Sharpness Ship Canal.

Maintenance yards

Each canal has its maintenance yard or yards, where the means of looking after the waterway are housed. These are distinctive places, with their workshops, perhaps dry-dock and the boats which need to be used. The one at Hartshill on the Coventry Canal is often photographed because of its pleasant buildings and layout, complete with clock. Another well-known one is at Bulbourne on the Grand Union, and this specialises in making lock-gates. View it from the window of the local pub. There's a pleasant small one near Fradley Junction, and a rather crowded one at Mirfield in Yorkshire. Mysterious places, some of these. But they are absolutely vital to the waterways, and of course their collections of working boats are not usually to be seen at home, but are scattered around. A visit to a yard – especially one making lock-gates for example, is an intriguing experience if it can be arranged.

Canal maintenance yard at Hartshill on the Coventry Canal.

Maintenance boats

The boats from maintenance yards are every-where – little "flats", bigger work-boats both narrow and broad, doing all sorts of jobs and carrying all sorts of equipment (since many tasks are only approachable by water). Dredgers are at work, and perhaps the most startling sights are the big dredgers on the Trent. They have cables out to the banks, and a white and a red indicator at their masts. You must pass on the side where the white indicator shows, or you may run foul of a cable. Needless to say, all working boats – especially on narrow canals – should be passed very slowly, as they may be doing a delicate job such as repairing the banks, which would be upset by a boat passing at speed. A cheery word to the men working in isolated places is often welcome.

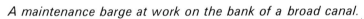

The smallest maintenance-boat — a "flat".

A maintenance barge at work on the bank of a broad canal.

Commercial boats

I personally get the greatest pleasure from seeing commercial traffic on waterways. After all, that was what they were built for, and on the whole the boatmen are kind to us newly-arrived pleasure-boaters. The rarest sight is a pair of working narrow boats, but there are still some around, usually run by real enthusiasts who don't expect to make a fortune. They carry whatever they can persuade people to send, but usually coal, bagging it up and selling it to houses along the waterway more cheaply than they can get it by land.

Most commerce, however, is not on the narrow canals, but on the wider ones mostly in Yorkshire, or on rivers. In Yorkshire, sadly, you'll no longer see the famous "Tom Puddings" in long trains, but the shorter sets of three compartment boats run backwards and forwards to power stations, pushed by their sturdy little tugs. You'll see many barges, too, on the Aire & Calder and the Sheffield & S. Yorkshire, carrying sand, coal, gravel, oil, etc., from and to Hull, Goole, power stations and the Trent. Gravel-barges load at wharves on the Trent, and sand-barges unload at Knottingley.

Ships may be an alarming sight to meet, but they are moving pretty slowly where you see them, such as on the Yorkshire Ouse, (watch the hairpin bends), on the Gloucester & Sharpness, on the R. Weaver and on the lower Trent. If one appears from behind you, turn around in good time and pass it in the opposite direction to meet its wash. Then turn round again. This applies to barges, too, if the waterway isn't very wide, since the wash may rock you.

Rare sight — a specially built narrow-beam boat for carrying pottery between factories on the Caldon Canal. There are far fewer breakages this way.

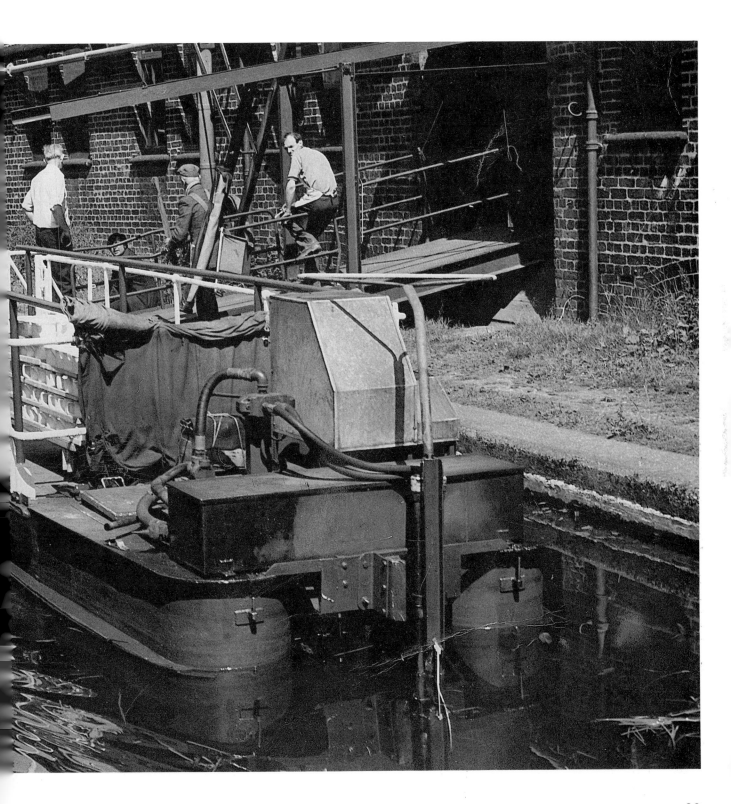

Nostalgic view of part of an 18-container "Tom Pudding" train, which carried coal from Yorkshire collieries to Goole until very recently. ▲

Modern tanker-barge running into Yorkshire, seen in an Aire & Calder lock. ▶

Facing page, top:
You'll meet ships on a few wide waterways. This one is passing through a swing-bridge at Selby on the Yorkshire Ouse.

Other boats

Beside private and hire cruisers, and the commercial and maintenance boats and ships already mentioned, there are still other types of vessels to be spotted.

A pair of camping boats — canvas-covered ex-working boats — entering the lock by the Waterways Museum at Stoke Bruerne — a popular place with school parties. ▼

You'll probably meet *canoes*, even, and certainly during the summer there will be *camping boats* filled with young people – some disciplined, some not. These are usually converted working narrow boats, with canvas covers and simple camping equipment, but others are more sophisticated. They are an excellent way of introducing youngsters to waterways.

Anyone who does not feel like doing the work of boating and locking can book a cabin on a *hotel-boat*, for there are quite a few of these roaming the system. Many are in pairs, like the old commercial "motor" and "butty", and although the cabins must be small they can be quite comfortable, with in some cases excellent meals. You can walk along the towpath or help with locks if you wish, or just have a lazy life.

To sample canals if you don't know them, there are dozens of *trip-boats* all over the system, doing short or longer trips during the day. Perhaps the most intriguing type of trip-boat is *horse-drawn* and indeed there are several of those scattered about. One runs from Froghall at the far end of the Caldon Canal and another travels near to the site of the Battle of Bosworth Field, on the Ashby Canal.

Then there are travelling art-galleries, puppet shows, theatre companies – you name it, somebody takes it about in a boat. There's even one showing things "Made in Britain", and I've heard of one selling computers. Keep your eyes open, for there's never a dull moment for the observant boater.

A pair of hotel-boats moored among other pleasure boats at Stourport, where the Staffs & Worcs Canal meets the R. Severn.

Trip boat with passengers on the Kennet & Avon Canal. It is in a curiously-shaped rebuilt lock which previously had attractive sloping turf sides. ◄

You can take a nostalgic short trip on a horse-drawn boat in a number of places. This one is on the Ashby Canal, near the site of the battle of Bosworth Field. ▼

Numbers and notices

All over waterways you'll find old notices and numbers. Numbers may seem unimportant items, but in fact, as I've mentioned elsewhere, the numbers on bridges especially are often the only way of knowing – with the aid of your guide – just where you are. Some of the old bridge-numbers remain, but others have sometimes been stolen as souvenirs. Newer ones have been put up, and on some canals, such as the Staffs and Worcs, there are striking bridge-names surrounding the numbers, kept up-to-date by the devoted Canal Society.

There are numbers on some locks, too, and you'll perhaps enjoy (or not?) following them up or down from 1 to 58 on the Worcester and Birmingham Canal. Some such numbers are carved heavily into stone, as in the locks at Wigan, or those at Wolverhampton. Notices vary from some unusual ones on the Shropshire Union and Llangollen Canals (about bridge loads, putting down lift-bridges, tipping stone, etc.) to those about not banging in mooring-pins where there are electric cables (Calder & Hebble, also Nottingham Canal). Some have been rescued and can be seen at the museums, including one threatening transportation for wrong-doers.

Mileposts are almost a study in themselves, for you can see strikingly the variations between waterways. The Trent & Mersey ones have all been revived by the Canals Society, and have their own distinctive shape, quite different from those on the Shropshire Union, for example. A few Grand Union ones remain of a different shape again, while the Leeds & Liverpool ones are more like some road mileposts. The Lancaster ones are oval and the solid stone Macclesfield posts have now been restored by yet another canal society.

Other striking posts are on the River Medway and the Gloucester & Sharpness Ship Canal, but you'll spot others here and there, as well as boundary posts marking the canal property.

Large numbers of canal bridges have numbers, but those on the Staffs & Worcs Canal have attractive plates with their names as well.

There are numbers on many locks, and the Wigan flight runs to Roman ones carved in the lock-shoulders.

Look out for old notices, especially on the Shropshire Union Canal.

Many canals have mile-posts in distinctive designs. This is on the Leicester line of the Grand Union (a part once called Grand Junction).

"Stop-planks" to dam off the canal if necessary. These are on the Bridgewater, needing a crane to lift them.

Stop-planks and stop-gates

Here and there along canals you'll notice a pile of planks, often near a bridge or lock. These are stop-planks, and you will see grooves in the concrete sides of the canal into which they can be slotted one at a time until they reach to the water-level. In fact this forms a dam to cut off the water, and with another lot of planks elsewhere, water can be pumped out of the length in between them.

One common use of such planks is to dam off a lock so that work can be done on it after the water is removed. Tunnels can be dammed off and emptied in the same way. And if a canal bank is damaged so that water leaks out, that part can be sealed off with stop-planks. Sometimes you see a single lock-gate – at each end of an embankment, for example. This is another form of "stop", which can be swung across to dam off the length concerned. Many such stop-gates have fallen into disuse now, and you'll see their remains.

A stop-gate instead of planks – at the end of a long and high embankment on the Shropshire Union Canal.

Junctions

Along roads you come to junctions all the time, with signposts to endless places. Along waterways junctions are normally something of a rarity, and often a surprise. They may be twenty or thirty miles apart, and indeed most of the Leeds & Liverpool runs for 92 miles between junctions, apart from the short branch at Skipton. The whole of the Kennet & Avon has no junctions now.

The BCN, in contrast, has junctions all over the place, many of them now leading nowhere, but still perhaps with a lovely BCN towpath bridge over what once led to a lost branch.

Thus a junction is often something to look forward to. A few are lonely, but most are busy with boats and people. Fradley Junction on the Trent & Mersey is one of the latter, with its famous *Swan* pub. Another is at Braunston, where all the boating world seems to meet at times. But the beginning of the Llangollen is very rural, like Duke's Cut Junction between the Oxford Canal and the Thames.

Torksey is intriguing, where you leave the Trent for the Fossdyke and Lincoln, and use unusual capstans to work the lock. The lock from the Stratford Canal to the Avon is also memorable, in front of the Theatre. Great Haywood has a photogenic towpath bridge, and the junction between the Trent & Mersey and the River Weaver at Anderton has of course the famous boat-lift. There's even a cross roads at Trent Lock, where the Soar comes in one way and the Erewash leaves the Trent opposite. There are more signposts than there used to be, and at one time many a beginner has been known to miss a vital junction on his route. But all these junctions have character, and are worth collecting.

Well-known Hawkesbury Junction, where the Oxford Canal leaves the Coventry. Once a busy halt for commercial narrow boats, this pub now serves pleasure-boats equally well.

Looking at Nature

Recently there has been antagonism from some selfish people, usually landowners, but also (sadly) some anglers, to the restoration to navigation of some waters. Hypocritically, pretending to use a "conservation" argument , they make the most outrageous accusations about the harm done to wildlife by the passage of boats. Yet anyone using a canal or river will see only too clearly the luscious growth alongside, and the many insects, birds and animals certainly not bothered by boats. The fact is that the restoration of what were dry or rubbish-filled old canals has provided a new vast linear "nature park" where one didn't exist. And as we cruise those waters we can see this very plainly indeed. The return of boats has meant the return of havens for living and growing things.

This isn't the book to have whole chapters on bird-watching or flower-identifying, and if these interest you, you'll bring your own books. But whether you are particularly keen or not, you'll see herons, kingfishers, moorhens, coots, grebes, ducks, Canada geese and swans galore, despite those many killed by lead weights and nylon fishing-lines. And of course more common small birds like the peace of waterway banks, especially wagtails. They all seem much tamer here than out in the other world, and many will come to be fed.

You'll see animals such as water-voles (wrongly called rats), and even the rarer badgers and foxes if you're lucky, with many a glimpse of rabbits and hares in the fields. In particular, you'll see a wealth of flowers that you would have to search for far and wide elsewhere and they grow huge, too, with the water supply always there. Small spring flowers alongside are delightfully visible before the grass grows, and yellow irises are common in the water, with even water-lilies in some areas.

Swans — so vulnerable to fishing-lines and the old lead weights — will often feed out of the hands of boaters.

Look out for "bulrushes", correctly called reedmace, among the less striking variety of reeds.

In autumn you can pick and eat blackberries and crab-apples (cook them together with plenty of sugar), and there are many bright berries also. In May there is a huge variety of fresh green that you won't see at other times, as it all seems to darken later.

There are many other things to see with the casual eyes you develop when slowly cruising, and I'll look at some particular items in the next chapter. But the guides, too, will tell you of historic mansions and other places to visit not too far from many a waterway.

11

SOME WATERWAY HIGHLIGHTS

After that general look at waterway surrounding, here are a few comments on some special places to be found along these canals and rivers. It isn't everybody who gets off a boat and spends a lot of time exploring the countryside, since this is done so much more easily by car. But near to our waterways, or actually alongside them, are many odd landmarks (watermarks?) worth mentioning and exploring. So here's a selection, but others will also be listed in the better guides.

Anderton Lift

You come across this amazing contraption between the Trent & Mersey Canal and the River Weaver down below. It's been in mechanical trouble a lot recently , and gave rise to a big campaign to keep it functioning. It carries boats up or down between the two waterways, floating in a tank of water. You might even think of it as a moving lock. It's an interesting experience, though umbrellas are recommended to catch any grease coming down from the mass of pulleys up above.

Barmby Lock

This lock with odd-shaped gates controls entry from the Yorkshire Ouse to the River Derwent. The Ouse is tidal, so you need care to go there.

The gates, controlled by a lock-keeper in a cabin up above, swivel to open. Their shape is due to the fact that sometimes the Ouse outside is higher than the Derwent inside, so ordinary "pointing" gates would swing open.

Barton Aqueduct

The historic Bridgewater Canal originally crossed the River Irwell on Brindley's brick-built aqueduct, which people thought a remarkable idea. When the Manchester Ship Canal took over much of the Irwell's route, there had to be clearance for big ships passing under the Bridgewater. So a swinging aqueduct was built. It can be opened, full of water but not boats, for a ship on the big canal below. The operation is fascinating to watch.

Bingley staircase

This set of five locks on the Leeds & Liverpool Canal draws big crowds in summer, to see boats climbing up the steep slope. Luckily for new boaters there is a helpful lock-keeper to control their antics, for it can get very confusing having to work things out in a crowd of spectators.

◄
The remarkable Anderton Lift — to carry boats
between the Trent & Mersey Canal and the R.
Weaver. This was closed for some years, but may
be restored by the time you read this.

Unusual lock-gates from the Yorkshire Ouse to the
R. Derwent at Barmby. They allow for water-levels
varying each side.

The unusual swing-bridge which takes the Bridgewater Canal over the Manchester Ship Canal. It is sealed off by gates — seen moving here — and swung open, with the water still in it, to allow ships to pass on the Ship Canal below.

Bottle kilns
In the Stoke-on-Trent area you still see those strangely-shaped kilns, built of brick but looking like bottles. Look out for them on the Trent & Mersey Canal and on the Caldon.

Burnley embarkment
Nearly a mile long, this high length of canal strides out across Burnley, so that you get a birds-eye view of the town.

Bratch locks
Neither flight nor staircase, these three old locks on the Staffs & Worcs Canal are a bit tricky to work. The pounds between them are too short for boats, but extend sideways under the towpath to hold the lock-water passed down. So they shouldn't be worked like a staircase, but on the other hand, you can't pass another boat in them, so there are sometimes problems.

Caen Hill flights (Devizes)
This magnificent flight of 29 locks on the Kennet & Avon has been one of the most spectacular examples of dogged restoration. The spaces between locks are so short that the pounds have to extend sideways, making a very striking sight climbing up the hill.

Cathedrals

It's surprising how many cathedrals are close to waterways. Peterborough is near to the River Nene, and York Minster is a short walk from the Yorkshire Ouse. Lincoln is a steep climb, though, from the Witham, but both Gloucester and Worcester are a little walk away from the Severn. Ely is the most spectacular to see, for it is visible for some distance from the Great Ouse. There are others near waterways, too, such as Blackburn, Liverpool, Coventry and Ripon.

Many cathedrals are near to waterways. This is Ely, by the Great Ouse.

Compartment boats

On Yorkshire waterways you can see modern oblong containers, heavily loaded with coal and being pushed by tugs to power stations. A sad loss, though, are the old "Tom Puddings" — smaller containers towed for over a century until recently, in trains of as many as eighteen. They went down to Goole for lifting up and tipping into ships for export.

Cranes

You'll find old waterway cranes in a number of places along waterways, at wharves where they were used to load and unload goods. One is preserved at Braunston, for example, another at Cheddleton on the Caldon Canal, and one at the interesting museum at Llangollen.

Crofton pumps

On the Kennet & Avon, at Crofton, there are two old steam engines lovingly restored and often "in steam". They originally pumped water up from a reservoir to the top level of the canal, but this is now done by electricity.

Denver Sluice

This historic barrier was built to keep the tides of the Great Ouse from flooding the low-lying Fens. It's a massive sight across the river, and controls how much water runs down (or up!). There's a lock for boats, and inland boats come from the Middle Level for about quarter-of-a-mile in tidal water (at the right state of the tide) to reach the non-tidal Ouse up to Ely, Cambridge, or Bedford.

Modern replacements for the old "Tom Pudding" compartment boats — three large containers pushed from colliery to power station in Yorkshire. They are lifted bodily from the water and the coal tipped on to conveyor belts.

Preserved crane at Norbury Junction on the Shropshire Union. There is a hire base, a pub, and a British Waterways yard here, but no longer a "junction", since the canal to Shrewsbury is derelict.

Dudley tunnels

A mixed story. Three tunnels into each other, and of different dates, were restored with rejoicing in 1973 to become the longest still-navigable "tunnel" in the system. No internal combustion engines were allowed because of poor ventilation and the roof was too low for many boats. But the tunnel became very popular for trips, in which the boat had to be "legged" before it went over to batteries. That is, passengers lay down on boards across the boat and "walked" their feet along the tunnel sides to move it along, just as the old boatmen did before engines arrived.

Yet later the tunnel was closed again. A new length of the tunnel-end warren was opened and extended, and now trips just go into there from the Black Country Museum and back again, with much gimmickry for tourists. The real canal tunnel remains closed.

Dundas aqueduct

Rennie's lovely stone aqueduct carries the Kennet & Avon Canal – as an S-bend – over the River Avon below, between Bradford on Avon and Bath. The Somerset Coal Canal comes in near to it – a curious waterway, long closed, but now re-opened for a short distance for moorings.

Foxton staircases

These locks, near Market Harborough, nearly break the staircase record, since there are no less than ten of them. But in fact there is a very short pound in the middle, thus making two staircases of five locks each. They are narrow, so can lead to chaos since boats can't pass each other except in the small middle pound. Fortunately there is usually official help to sort boats out. There are always many gongoozlers around at weekends.

The water between locks uses long, narrow side-ponds, and the remains of a historic inclined plane, which once took boats up and down in tanks on rails, are being restored alongside.

Galton Bridge and cutting

On the main line of the BCN, the bridge is an ancient monument, a remarkable Telford 150ft arch of cast iron. The whole cutting which it crosses was a feat of engineering, lowering the canal to avoid the locks of the higher level. The view of the bridge has been spoilt by a new "tunnel" recently made by covering a huge concrete cylinder with earth for a new road.

Famous staircases at Foxton — two sets of five locks with a short gap between them.

Galton bridge over the Birmingham-Wolverhampton "main line". The "tunnel" in the distance had just been constructed when the photograph was taken, and is in fact a concrete tube covered with soil for a road.

Lea Wood steam-operated pump on the isolated Cromford Canal. It works regularly, and can be visited by horse-drawn trip-boat.

Some inland waterway! But famous Loch Ness is part of the trans-Scotland Caledonian Canal.

Harecastle tunnels

At a historic waterway spot near Stoke-on-Trent, Brindley's wonderful old tunnel has now sunk almost out of sight, but Telford's successor remains – though it, too, was almost abandoned a few years ago. After much restoration it now carries one-way traffic, controlled either by a tunnel-keeper or by timing notices. The old towpath used to make for a hazardous passage, but it has now been removed and the trip is less harassing. It's still 2936yds, though.

Hillmorton masts

This extensive collection of radio masts can be seen both from the northern Oxford Canal and from the Grand Union Leicester line. It's the Post Office (or no doubt British Telecom now) radio station for overseas communication.

Kings Norton guillotine lock

This unusual and unexpected structure, at the northern end of the Stratford Canal, isn't in use now. It was one of the old "stop locks" which shut off one canal from another to make sure no water was lost. They also acted as toll-taking places for the different canal companies. This one has a guillotine gate at one end, now kept up, and is much subject to photography.

Lea Wood Pump House

Although you can't cruise to the open part of the Cromford Canal near Cromford itself, you can drive there and take a trip on a horse-drawn boat. This will carry you gently to the Lea Wood Pump House with its tall chimney. And if you pick the right week-end you'll actually find it steaming away, and be able to see how this happens. You'll enjoy the canal's magnificent scenery, and the pump house is certainly worth a visit.

Loch Ness

Well, you have to get yourself up to the Caledonian canal first, but there are plenty of boats for hire there, though not "canal-shaped''. Loch Ness, as mentioned in Chapter 12, forms a large part of the canal, but no view of the monster is guaranteed,

Motorway bridges
In complete contrast to the whole canal atmosphere, motorway bridges have certainly made an impact on waterways everywhere, Some are more like wide tunnels, and have no beauty compared with the old canal bridges. But there's a remarkably peaceful atmosphere about some of them, since you hear (and often see) nothing of the traffic above. Among the Birmingham canals especially you can even moor peacefully under motorways.

Museums
There are several museums around the waterways now, and many small exhibitions in different places. For details see chapter 14.

Newark Castle
Rising up by the very side of the Trent, this old castle had King John die in it, and was knocked about in the Civil War. There's also a fine old bridge nearby, which takes the original Great North Road over the river.

Pontcysyllte aqueduct
Already mentioned earlier in connection with crossing it, no book on canals fails to extol this vast landmark near Llangollen, built of 418 iron plates to make its trough, and standing on 18 pillars, 121ft above the Dee at one point. The view down as you cross is alarming, but it's also worthwhile climbing down to see it from below.

Power stations
Like motorway bridges, these have come to make their modern mark on inland cruising. I find them fascinating from afar and you see several from the distance down the Trent especially. There are many others throughout

the system. Some of those in Yorkshire have their coal delivered by barge or pushed containers and the one at Ferrybridge often looks eerie as you approach it in certain weather conditions.

Shakespeare Memorial Theatre
As you emerge from the tunnel-like road-bridge at the southern end of the Stratford Canal, there by the side of Bancroft Basin is the Theatre. You can visit it from the basin, or (if you have an Avon licence) you can moor in the river opposite.

Sheffield Basin
I regard this as a necessary pilgrimage for inland cruising enthusiasts. This isn't for any striking reason, but because it's one of those places that people forget about, or think too remote to bother with. Yet a trip there is well worth while, for you now have new large automatic locks until Rotherham. But after that the locks give you some strenuous exercise climbing up to the historic basin at the end. This, however, after a lot of argy-bargy, is being ''developed'', so see what you think of it when it's finished.

Stanley Ferry aqueducts
These are a memorable pair on the Wakefield Branch of the Aire & Calder. The striking original wrought iron trough, suspended from cast iron arches, is an ancient monument now closed to boats, and a new (ugly?) concrete one is in use alongside. This, unbelievably, was constructed on land and pushed across, inch by inch, in 1981.

Steward aqueduct
There are two ''main lines'' for part of the way from Birmingham to Wolverhampton, and the ''new'' part was cut lower than old. At one place the old actually now crosses the new line on the Steward (or Stewart) aqueduct. It's rather swamped nowadays by the M5, which

Power stations often offer striking views, such as this one on the Aire & Calder Navigation at Ferrybridge.

Three forms of transport near Birmingham — M5 above, canal below, and railway alongside. The low brick structure is Steward Aqueduct, carrying the old main line of the canal over the new.

crosses higher up at almost the same spot. As an historic oddment, there's also a railway alongside the new line, so you have potted transport history at one place.

Stretham pumping engine
On the Old West length of the Great Ouse you'll find this historic one-time method of draining the surrounding area. It has a Boulton & Watt beam engine with a huge scoop wheel. It took the place of four of the ubiquitous fenland windmills, and can lift 30,000 gallons of water a minute when working, using a ton of coal every six hours. Although now itself replaced with modern pumps, it is still intact, and you can call and visit it.

Wigan
Once a joke, Wigan has triumphantly staked a claim as a "must" for boaters. Maybe not for its 23 locks, though many enthusiasts travel far to enjoy the battle up the hill, with the unusual paddle-gear (disappearing fast?). But Wigan now has a very visitable "Wigan Pier" complex at the bottom of the flight, where the attractions include an excellent view of life in the not-long-ago past.

York
The whole place is a landmark for boaters, who will have travelled the tidal way up the Ouse, through Naburn locks and past the Archbishop's Palace to this delightful city. The Viking Hotel stands straight out of the riverside and you can moor further on by a tree-lined stretch and walk a mere few yards to the Minister. Then there are the Vikings at one end of history, the Railway Museum at the other, and, as they say, "much, much, more".

The Zoo
Amusingly, the Regents Canal actually runs through part of the London Zoo, and on this pleasant tree-lined stretch there are quite a few inhabitants on view. If London is too far south for you, you can visit Chester Zoo, half-a-mile from Bridge 134 on the Shropshire Union Canal.

Brave the tidal Yorkshire Ouse to pass Naburn locks and enter York — a magnificent place to reach, and you can moor near the Minster.

Winter view down to Napton Flight on the Oxford
Inset *Frost on the roof above Audlem*

In the heart of the Birmingham Canals ▲
Lost in a lock at Bratch, near Wolverhampton ▼

Dropping down to Stratford-upon-Avon

Canoes below a weir on the Chelmer & Blackwater

Carrying tractors
for Africa ▲

Beautiful rope-weave on
a butty helm ▲

Sea-going traffic on the
Yorkshire Ouse ◄

12

YES,
BUT WHICH WATERWAY?

It's all very well my going through the business of using a boat and going on canals and rivers, but where will you choose to go?

This is the 64,000-dollar question, for our waterways have such wide variety, and people's tastes differ so much, that I can't answer such a question for *you*. All I can manage is to give you brief descriptions of the waterways on offer, and leave it to you. I'll try to do this in as potted a way as possible without being useless. Even then, my descriptions are only my personal views, and others will have other opinions. Moreover, I describe each waterway as a whole under its name (happily the old names are still in use). But most people when cruising visit bits and pieces of different canals or rivers, since they're all linked up at scores of junctions.

So do read what I say in conjunction with the small map on p.130, or better still buy a bigger map (see p.156), to think of planning a journey. And of course, if you are hiring a boat, you need brochures from some of the firms mentioned on p.164 to discover where you can start from.

Two other things should be considered: If you haven't a fairly energetic crew, then don't go for some of the canals – such as parts of the Leeds & Liverpool, or the northern Grand Union and other canals which head for Birmingham – where there are whole flights of locks. Look for areas where there are fewer locks. Again, not everyone enjoys going through long tunnels, so look out for those, too. It's usually possible to walk over the top, however, and I have known nervous passengers ring for a taxi and then wait at the other end.

If you become really interested in the whole waterway network, there is a wonderful "bible" available which tells you every tiny detail of every navigable waterway – and some unnavigable ones too. This is *Inland Waterways of Great Britain*, a labour of love by L.A. Edwards (Imray Laurie Norie & Wilson). It's essential reading for even the mildest enthusiast – a treasure of facts, distances, locks, bridge-heights, maps, and a great deal more (see "Books" in Chapter 13).

So here goes with my own highly-potted effort – waterways in alphabetical order, and my own views on them. I must say that I haven't just lifted all this from other books, as is so often done. I have cruised on almost every inch of the navigable waterways I mention, as well as on parts of those being restored. So whether you agree with me or not, at least it's all first-hand! But for every last detail of the routes, you must have the guides mentioned in the next chapter.

N.B. The number by each waterway will help you to find it on the map on p.130.

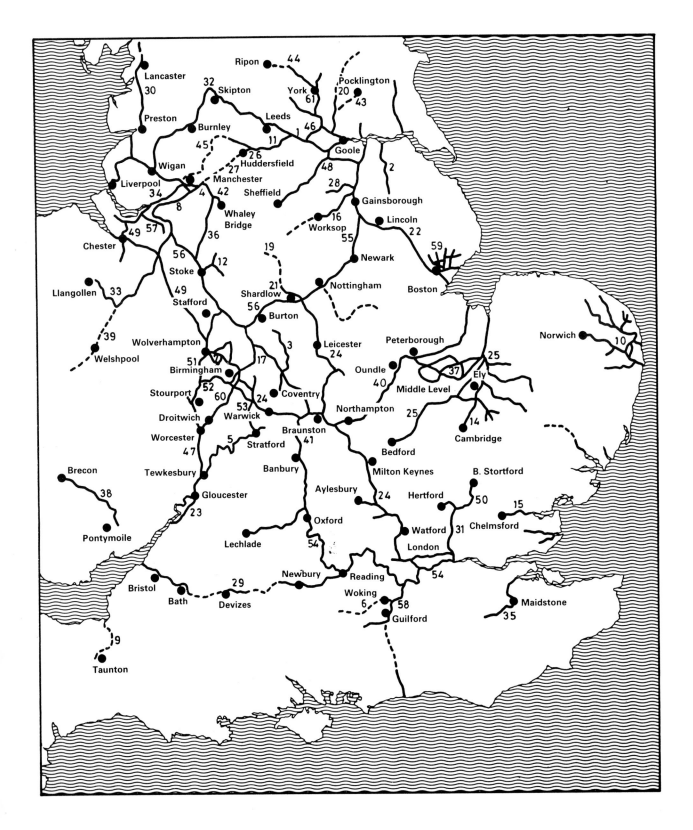

Facing page:

Our inland waterway system. The numbers on waterways help you to identify them in the list in the following chapter. Obviously you'll need a much larger map for more details, and some are mentioned in Chapter 13.

1 Aire & Calder Navigation

A wide, deep commercial waterway in Yorkshire, busy at times with both large barges and "push-tows" of three containers full of coal for power-stations (or running back empty). Their steerers are helpful, but of course can't work miracles, so keep out of their way. Locks are big, have traffic-lights, and are worked by keepers in cabins. Obey what they say. I find the locks almost more peaceful to go through than some smaller ones.

Finding a mooring isn't easy, since you can't just tie up at the side as elsewhere, or passing barges will pull out your mooring-pins as their powerful wash moves you along. Ask lock-keepers for suggestions, as there are several alongside moorings provided for pleasure boats. Don't be nervous about cruising here. It can be especially enjoyable, as the width allows even the steerer to relax and look around. You can get to Leeds and Wakefield on the A & C or its branch, and to the Calder & Hebble Navigation. There's also a junction with the R. Aire leading to the Selby Canal, and thus to Selby and the Yorkshire Ouse.

(From Goole on the Yorkshire Ouse to the Leeds & Liverpool at Leeds, with branches to Wakefield and Selby. 34 miles, 13 big locks to Leeds. 7 miles, 4 big locks to Wakefield. 12 miles, 4 broad locks to Selby).

2 R. Ancholme

This river is not normally approached by inland boats from elsewhere, since the journey involves the tidal and risky Humber. But you can trail a boat there, and quite a number are kept there.

It looks like a ruler on the map, and runs straight for most of its way, with only Brigg of any size on it. There is a good pub at Brandy Wharf. The one unusual lock on the river is often out of use, but there are only about two miles cruisable above it.

(From R. Humber at Ferriby to Bishopbridge. 19 miles, 2 broad locks including sea-lock).

3 Ashby Canal

The Ashby's claim to fame is that it hasn't any locks. And there are also some long lockless lengths leading to it. So lock-dislikers love it.

It doesn't go to Ashby (de-la-Zouch), and never did quite manage to. It's also shorter now than originally. But you still have 22 miles of rather shallow water, past Hinckley and some pleasant villages, one of which has a railway museum and steam trains running. There's also a horse-drawn trip boat and elaborate displays about the Battle of Bosworth Field.

(From near Bedworth on the Coventry Canal to the terminus south of Measham, 22 miles, no locks).

4 Ashton Canal

This is a complete contrast to the Ashby, being entirely within the Greater Manchester built-up area. It's not a good place to stay the night, but makes for a memorable cruise with many locks. Its recent history is interesting, for it was derelict and in danger of vanishing entirely. But powerful campaigning – and hardworking voluntary clearances – turned opinions right round, so that it was restored to full use. It now forms part of the so-called "Cheshire Ring" of both rural and urban canals.

(From the Peak Forest Canal at Ashton-under-Lyne to the Rochdale Canal in Manchester. 6 miles, 18 narrow locks).

5 R. Avon (Warwickshire)

A stirring recent history, in that it was derelict for some time until one of the earliest restoration campaigns restored it from the Severn up to Evesham. Then another restored the rest to above Stratford (with plans for a future link with the Grand Union at Warwick). Now it offers a delightful cruise amid orchards and towns such as Evesham, Pershore and Stratford itself. Separate licence fees, and mooring isn't always easy.

(From Tewkesbury on the Severn to (and above) the Stratford Canal at Stratford. 43 miles up to Stratford, 17 broad locks).

6   Basingstoke Canal

At the time of writing the canal is not fully open, but may well be as you read this. It is now owned by Surrey and Hampshire, and its restoration was really brought about by vigorous campaigning and work from an excellent Canal Society. In this way its many locks have been restored and the channel dredged by large numbers of volunteers. Boats are now returning led by a profitable trip boat, the *John Pinkerton.* Although seemingly surrounded by built-up areas, it is in fact quite remote in parts, running near much military land and camps. At the moment the navigable length ends at Greywell tunnel which is impassable (and enjoyed by bats). All the locks are in the first 16 miles, with a further 15 miles level.

(From the R. Wey near Byfleet to Greywell tunnel, 5 miles short of Basingstoke. 31 miles, 29 broad locks).

7   Birmingham Canal Navigations

This warren of waterways could take a book to itself to describe. But briefly, a variety of canal companies produced a vast network in the Black Country area around Birmingham, Wal-

Peaceful scene on the restored Warwickshire Avon.

132

sall, Wolverhampton and Dudley, and then amalgamated to become the "BCN". They served the factories of this industrial area, and had hundreds of short branches for this purpose. Many miles have been lost now, but there are still over 100 navigable miles for you with well over 100 locks, at three slightly different levels. With the whole area being on a plateau, it means that the six different approaches from the rest of the system all have to climb a goodly number of locks.

Cruising in the BCN has its fascinations, but its hazards as well. "Yobboes" can be a nuisance, especially in school holidays, and there's more rubbish around than usual to get on your propeller. But many enthusiasts thoroughly enjoy the remarkable views and surroundings in parts of this unique area.

Most people, unfortunately, merely go from Birmingham to Wolverhampton on the "Main Line", to get from one rural area of canals to another the other side of the Black Country. This is a pity, for even the "Old Main Line" covering part of the same route is worth the detour. But also the Wyrley & Essington (its "Curly Wyrley" nickname reminds you how to pronounce it), running along the north of the area, is a worthwhile trip. Or you can go south of the main lines to try the wide, two-towpath Netherton tunnel, and then down the Stourbridge flight to the Staffs and Worcs Canal. Throughout the BCN there are in fact seven tunnels, though two of them are short, recent, and artificially made ones. It's a pity, too, that the through part of historic Dudley tunnel is still closed. A small section at the northern end can be entered by trip-boat, and the excellent Black Country Museum is near there. There's much else in the warren of the BCN to attract the adventurous.

(All over the place in Birmingham, Perry Barr, Brownhills, Walsall, Wolverhampton, Wednesbury, West Bromwich, Smethwick, Oldbury, Dudley, Tipton, Brierley Hill, etc. 106 miles, 139 narrow locks, give or take).

8 Bridgewater Canal

Usually named as the "first" canal, this isn't actually true. The Romans built our first one, and there were other canal lengths of sorts before the Bridegwater. But it was a true canal, the beginner to the present canal system. Built by Brindley and Gilbert for the Duke of Bridgewater, it took his coal from the Worsley mines (with their own many underground canals) more cheaply to Manchester. The masonry Barton aqueduct which took the canal over the Irwell was a miracle of its time. It has since been replaced, when the Manchester Ship Canal was built beneath, by an even more remarkable swing aqueduct (see p.109).

The canal was also extended across Cheshire to Runcorn, where it locked down to the Mersey (later to the Ship Canal). But the locks, sadly, are now gone. The Trent & Mersey Canal comes north to join it at Preston Brook, so you can then amble along its long lockless length either to Manchester or further north (via its Stretford & Leigh Branch) to the great Leeds & Liverpool. But pop also down the 5-mile bit, now like a mere branch, to Runcorn, where you get an especially good view of the Ship Canal.

(From the Rochdale Canal in Manchester to Runcorn, 28 miles, no locks – short branch to the Trent & Mersey Canal after 23 miles of this. Branch from Stretford to the Leigh Branch of the Leeds and Liverpool, 11 miles, no locks).

9 Bridgwater & Taunton Canal

Another canal which is not linked to the network, but is being restored with the help of enthusiasts. It runs through pleasant Somerset countryside, and those with boats that can be trailed would enjoy a visit, as the movable bridges are being restored and the locks are re-opened. Some locks have unusual paddle-gear.

(From Taunton to Bridgwater. 14 miles, 6 broad locks).

10 The Broads ("Norfolk & Suffolk Broads")

This is a distinct area of waters separate from the main system, and popular for pleasure cruising long before canals were "discovered" for this purpose. It is a mixture of large lakes (the "broads") and connecting rivers. Its atmosphere is entirely different from elsewhere, perhaps a bit like the Thames though with no locks and more hire-cruisers. The highly organised hire industry offers boats quite unlike "canal" shapes, from many bases. The whole area is geared for boating, and there are pubs and villages everywhere offering facilities.

You can even cruise into the heart of Norwich, as well as near to Lowestoft and less easily near to Yarmouth (where the tide makes itself felt more strongly). There are still quite a few sailing boats, though fewer than at one time. It is a startling sight to see a tall sail making its way up the narrow rivers against the wind. But somehow the many motorised boats seem to get on with the sailing ones. It is fair to say that the Broads were the birthplace of boats hired for pleasure on any large scale.

(Mostly in Norfolk, but a small part in Suffolk. About 115 miles, no locks).

11 Calder & Hebble Navigation

This, I often feel, is my favourite, for no particular reason. Maybe it's because on the map it looks to be all along industrial lengths, but in fact you're surprisingly sheltered from them. There's much grass and many trees, and I've gathered blackberries by some locks. Yet you can step off your boat roof into the market at Brighouse.

There's a short branch to a basin near Dewsbury, and a link with the Huddersfield Broad Canal and thus to the Narrow Canal now being restored.

You can, helpfully, hire boats at the top end at Sowerby Bridge, or reach the navigation via the commercial Aire & Calder. It's a mixture of river and canal, in fact, with purely canal at the top end reaching to the Pennines towering above. The rapidly restoring Rochdale Canal leaves it there.

It has many interesting aspects, not least the wooden "handspikes" to open much of the paddle-gear. And its locks vary in size and mechanism. The basins at both Sowerby Bridge and Brighouse are picturesque.

(From the Aire & Calder Branch at Wakefield to the Rochdale Canal at Sowerby Bridge. 22 miles, 27 broad locks plus flood locks. Branch to near Dewsbury, about 1 mile).

12 Caldon Canal

Really a branch of the Trent & Mersey, this goes delightfully into the hills from Stoke-on-Trent. The upper end especially is almost secret, running in the Churnet valley (part of it using the river), and then to Froghall with its low tunnel and horse-drawn trip-boat. There's a branch towards Leek, which leaves the main canal and passes over it after the main line has dropped through three locks. This branch doesn't reach Leek now, but after an unusual pool and a short tunnel, ends abruptly. It's worth visiting, though, for the views alone.

(From Etruria, Stoke-on-Trent, to Froghall. 17 miles, 17 narrow locks. Leek Branch, 3 miles, no locks).

13 Caledonian Canal

Quite different from any other waterway, this mighty cross-Scotland route is a mixture of lochs and locks, huge natural lakes with canal cuts between them. One of the lochs is Loch Ness, which contributes 25 miles to the "canal". The locks will take boats 150ft long or more, and 35ft wide. Thus, besides many pleasure-craft, fishing-boats and other big vessels use the waterway. You can hire boats on the "Cally", but they are sea-going cruisers rather than the stolid narrow boats of canals to the south. And of course they need to be, as the lochs can be quite sea-like in stormy weather.

(From near Inverness to near Fort William. 60 miles, 4 lochs, 29 big locks).

Fishing boat crossing Scotland via one of the locks in the Caledonian Canal.

14 R. Cam – see R. Great Ouse

15 Chelmer & Blackwater Navigation

Remote from the network, this is a shallow and rather unknown river-based waterway, running from Chelmsford to tidal waters at Heybridge. It is privately-owned still, and use is limited. It is a peaceful route, worth developing.

(From Chelmsford to Heybridge, 14 miles, 13 broad locks).

16 Chesterfield Canal

My home canal – well, I was born near it, fell into it, and walked along its ice when a lad. But even falling over backwards to be fair, I still think it's an unusual canal. For a start, it doesn't link up to other canals, but has to be approached with caution via the tidal Trent, which means that many people don't visit it. Thus you feel it's quite an adventure even to get on it. And certainly you must take advice about the best state of the tide to set out for, and arrive at, the entrance lock at West Stockwith. But so long as you've warned the lock-keeper of your coming, he'll be there to help you in.

YES, BUT WHICH WATERWAY?

Retford Basin on the Chesterfield Canal, which is only navigable to Worksop at the moment. ◀

Once in, it's truly rural most of the way, first through broad locks including one called Whitsunday Pie, then a delightful market town at Retford, and narrow locks from then on through "Dukeries" surroundings to Worksop, the present terminus. Whether the canal will ever go to Chesterfield again is a matter for campaigning, for among other things there's a brute of a tunnel to restore. But the Canal Society is a lively one, which never gives up.

(From the Trent at West Stockwith to Worksop now. 26 miles, 16 locks, mostly narrow).

17 Coventry Canal

Most people don't actually visit the start of this canal in Coventry, but join it about five miles out on their way up north. This is a pity, for a lively Canal Society has helped keep the route to the city pleasant, and the canal basin there is well worth visiting, handy for city and cathedral. From Hawkesbury Junction where the Oxford comes in, there's a fair amount of built-up area through Bedworth then Nuneaton – town of many allotments, it seems – before more open country. Atherstone, with a flight of locks cut off from the town by the railway, offers shops, then it's quite enjoyably rural and level again until the outskirts of Tamworth. There are two locks here, and a lovely little aqueduct over the Tame. Then it's 11 winding miles, occasionally seeing the A38, to historic Fradley Junction with the Trent & Mersey, much photographed pub, boats and all.

(From Coventry to the Trent & Mersey at Fradley. 38 miles, 13 narrow locks).

18 Crinan Canal

Scotland's only other fully-open canal besides the Caledonian (though others are partly in re-use), this is an isolated short canal with a striking character of its own. It in fact cuts across the neck of the long peninsula of the

Mull of Kintyre, giving boats 9 miles and 15 locks to pass, instead of 132 miles round the Mull. Thus pleasure yachts and fishing boats save a long and maybe rough journey to the Isles.

(From Ardrishaig to Crinan. 9 miles, 13 big locks).

19 Cromford Canal

Apart from a short bit at the top of the Erewash, the only other part of this canal still open is quite isolated from the rest of the system. So you can't reach it by boat. But it's worth listing if only because of its fine scenery and the dedication of the Stoker family and the Cromford Canal Society. They have kept open a length near Cromford itself, and their horse-drawn trip-boat is very popular. It runs to the Lea Wood steam pumping house, which is regularly in use. The surroundings are spectacular. The canal originally had a link by rail over the Derbyshire backbone to the Peak Forest Canal.

20 R. Derwent (Yorkshire)

Cause of much argument about whether there is a right of navigation on its whole length, this river is quite delightful, and was certainly navigated up to Malton and beyond in the past. It is not often visited by boats from the canal network (and some hire firms won't let you try it) since it has to be reached via the tidal Yorkshire Ouse. If you are experienced enough to deal with the tides, you enter through a most unusual lock whose curved gates revolve. The lock-keeper will take your fee and lend you a key for the next lock. It's a lonely trip, with few villages. Most people pause at Sutton (or Elvington) lock, which has a guillotine gate, to visit Sutton on one bank or Elvington on the other. At the moment the terminus is at Stamford Bridge, with an unusable lock in which at times you can moor. Moorings are difficult along the river, as landowners here-abouts seem very selfish.

(From the Yorkshire Ouse at Barmby to Stamford Bridge and on to Malton. 22 miles to Stamford Bridge, 2 broad locks).

21 Erewash Canal

This canal, once continuing northwards via the Cromford and a railway to the north-western canals, is now rather out on a limb. It goes 15 miles north from the Trent opposite where the R. Soar comes in, and is mostly through industrial scenery.

Past a pub by the entrance lock – oddly called the *Steamboat* – you cruise through Long Eaton, some once-much-busier ironworks, Ilkeston, and to a pleasant basin (really the Cromford Canal by now) restored by enthusiasts. Here, as well as the Cromford, the derelict Nottingham Canal once came, and the lost Nutbrook Canal joined the Erewash further down. The Derby Canal, also, once left the Erewash at Sandiacre, so it used to be part of quite a little canal web.

(From the Trent near Long Eaton to Langley Mill. 12 miles, 15 broad locks).

22 Fossdyke Canal and River Witham

These two waterways form a continuous trip from the Trent down to Boston, via Lincoln. Apart from Lincoln, it's a very remote journey, first on an ancient Roman canal, then on a slow-moving river. Lincoln Cathedral is in view for some distance, and in the city you pass between Woolworths and Marks & Spencers. The Witham joins here, and after a lock with a guillotine gate you're out in the country again quite soon. There is only one more lock before the tidal lock at Boston, and long empty stretches to pass. There's a junction with the Slea Navigation which is being restored, and another at oddly-named Anton's Gowt lock with the Witham Navigable Drains which are weird waterways.

(From Torksey, R. Trent, to Boston. 48 miles, 2 broad locks).

23 Gloucester & Sharpness Ship Canal

Not the place for a week's cruise, but a very interesting section to visit if you are on the Worcester & Birmingham Canal or the Severn or Avon. You reach it down the Severn and through Gloucester lock, approached with care. This brings you through the docks to the Ship Canal which was dug to avoid some treacherous tidal reaches of the lower Severn. It has no locks except at its ends, and joins the river again at Sharpness, but that part of the Severn is not for inland pleasure boats. You may meet ships, which look alarming but aren't, since they crawl along. They just fit the many opening bridges, worked by bridge-keepers often living in charming little houses. Almost all have to be opened for pleasure-boats, too, and traffic-lights tell you when to proceed. Gloucester Docks has the National Waterways Museum, and near one of the swing-bridges you can visit Slimbridge Wildfowl Trust.

(From Gloucester to Sharpness. 16 miles, no locks except for entry, 16 movable bridges).

24 Grand Union Canal

This ought to read "Grand Union Canals", since the name is now given to an amalgamation of canals made in 1929. Before that they had several different names. I'll divide them into two main sections, though, and mention the branches of each.

London to Birmingham

In London itself you need a map to sort the GU out. There's a short Hertford Union section from the Regents Canal to the R. Lee, with three locks, and the Regents runs from Limehouse Basin off the Thames, through the Zoo, and to Paddington, with its famous "Little Venice" area. From there it is the "Paddington Arm" to the "Main Line" at Bulls Bridge, Southall.

This "Main Line" leaves the Thames at Brentford, running up locks on its long trip into the countryside, with a short branch to Slough on the way. From Denham onwards it's mostly greenery all the way as the canal begins to climb near the Chilterns. There's an arm to Wendover being restored, and a good navigable one to the heart of Aylesbury. Then through Milton Keynes and on to Stoke Bruerne with its museum and the mighty Blisworth tunnel beyond.

There's a branch with 17 narrow locks down to Northampton and the R. Nene, then another big tunnel, Braunston, and historic Braunston village. After running jointly with the Oxford Canal for five miles, the GU then starts on some new-type locks built in the 'thirties — masses of them as it heads for Birmingham past Warwick and Leamington. The last few miles are rather grubby, with narrow locks, along the edge of Birmingham to end under the present motorway "Spaghetti Junction". There are several links to the Birmingham Canal Navigations themselves.

Leicester Line

This section of the Grand Union leaves the Birmingham Line between Blisworth and Braunston tunnels, and was several different canals once (including a "Grand Union"). Although almost all the present Grand Union system has broad locks — except for some branches — there is a bottleneck in the first 23 miles here, with two groups of narrow locks. Thus barges could never pass north. Watford (not the Herts one) has some staircase locks, then there's a 21-mile level length in deepest country, to the famous pair of five-each staircase locks at Foxton. A branch at the bottom goes to Market Harborough, and the main line carries on through haunted Saddington tunnel towards Leicester.

There it uses the R. Soar, through Loughborough and at last to the R. Trent. In fact the Erewash Canal (q.v.), which leaves the Trent opposite, is also officially part of the Grand Union network.

(From London to Birmingham, and through Leicester to the Trent. About 250 miles with 274 locks, mostly broad).

25 R. Great Ouse and tributaries

The Great Ouse offers quite a small network of cruising waters. It can be approached from other inland waterways via the R. Nene and the Middle Level, with a short tidal length to Denver Sluice. From there the river runs at first between high banks towards Ely, its cathedral magnificently viewed from afar. The banks are lower then, and about four miles above Ely you can enter the Cam at Pope's Corner and visit the heart of Cambridge.

The Ouse itself goes from Pope's Corner via the Old West River through some delightful towns such as St Ives and Huntingdon, and eventually to Bedford. The locks mostly have guillotine gates at one end. From the Cam, you can enter some short "lodes", and from the Ouse there are three pleasantly navigable tributaries – the Wissey, the Little Ouse and the Lark. There's a link on tidal water from below Denver Sluice via the straight new Bedford River, which joins the above route at the end of the Old West. You must have an Anglian Water Authority licence.

(Non-tidal part runs from Denver Sluice to Bedford, with the Cam and other tributaries and lodes. About 123 miles altogether, plus tidal New Bedford 20 miles. 23 locks altogether including Denver. Some locks are given as only 10ft 6ins width).

26 Huddersfield Broad Canal

This is a short link from the Calder & Hebble to the heart of Huddersfield. Sometimes called "Sir John Ramsden's", it goes through 9 broad locks and under a unique lifting bridge to a basin. There's a vast American-style restaurant recently opened there, and a newly-dug link under a busy road to the Huddersfield Narrow Canal.

(From Cooper Bridge to Huddersfield. 3 miles, 9 broad locks).

27 Huddersfield Narrow Canal

This is a once-derelict canal over the Pennines, now being rapidly restored. Lengths of it are already open, with trip-boats operating, and an active Canal Society co-operates with sympathetic local councils to press on with the work. The greatest problem is the longest canal tunnel in the country, Standedge at 5,698yds. Its reopening will be a marvellous feat if it comes about.

(From Huddersfield to Ashton-under-Lyne. 20 miles, 74 narrow locks).

28 River Idle

Not often used, and special arrangements have to be made to enter it. But it is navigable by inland boats (without too deep a draught) from the Trent at W. Stockwith to Bawtry. Special big entrance sluices have to be raised at the right state of the Trent's tide. It's a lonely river, with only Misson village before Bawtry.

(From W. Stockwith on the R. Trent to Bawtry. 8 miles, entrance tidal sluices).

29 Kennet & Avon Canal

Perhaps this should more correctly be called "Navigation", since it consists of two rivers joined by a canal. Its great fame is the restoration feat carried out by the Canal Trust, which has worked for many years to revive this important and striking cross-England waterway. It links the Thames at Reading to the Avon and thus the Severn via Bath and Bristol, and was almost unusable for many years. Now it bustles with activity after a long slog. First it is the R. Kennet through Reading and on to Newbury, where the purely canal section begins. There's lovely country, though with plenty of locks and Crofton Steam Pump, through Pewsey and on to Devizes. Here a famous flight of 29 locks drops down in spectacular fashion, with extended side-ponds to take the water between locks. Again delightful countryside and on to Bradford-on-Avon and eventually Bath. There are several striking features such as Avoncliffe and Dundas aqueducts, and Claverton water-driven pump. At Bath the waterway joins the R. Avon and runs on to Bristol. Here inland pleasure boaters need

A cruise on the R. Idle, rarely visited by boats, but beautifully rural.

special care and knowledge to venture any further towards the Severn estuary.

(From Reading to Bath and Bristol. 86 miles, 104 broad locks).

30 Lancaster Canal

One of the few waterways not connected to the rest (except via a tidal river unsuitable for most inland boats). It's also unusual in having no locks on its main line, though there are six on a branch down to Glasson basin, with another out to the R. Lune. It starts unobtrusively in Preston, from where it was originally meant to cross the Ribble to join the Leeds and Liverpool Canal, but only a horse tramroad was ever built. Now the isolated canal runs up through pleasant country, with occasional basins, boat yards and small towns, to Lancaster, which now makes more of it than once. It almost touches the sea at one place, and the Lake District is in view (not to mention the M6) before a sudden end at Tewitfield. The rest of the way to Kendal was cut off when the motorway was built in 1955.

(From Preston to Tewitfield. 45 miles (including Glasson Branch), 6 broad locks on the branch).

31 Rs. Lee and Stort

These two rivers offer enjoyable cruising from London's East End into Hertfordshire, though there's much built-up area along them. The Lee (or Lea) joins the Thames at Bow, but you can reach it from the Grand Union via the Zoo and the Hertford Union. It takes you through Tottenham and Enfield to more rural areas via Cheshunt and Broxbourne. Eventually it comes to its cruising end in Hertford itself. The Stort takes you off near Hoddesdon, and is a narrower, more canal-like river. By Harlow new town it passes Sawbridgeworth to reach Bishops Stortford, with supermarkets near your mooring.

(R. Lee 28 miles, with 18 broad locks. R. Stort 14 miles, with 15 locks, given as 13ft 4ins width).

32 Leeds & Liverpool Canal

This is one of the mightiest canals, crossing the Pennines with magnificent views. It passes through several industrial towns but, like other canals, often unobtrusively. And the moorland scenery makes up for the industrial lengths. It isn't for the weak, since there are many locks quite heavy to operate, and over 50 swing bridges. Its paddle-gear is still among the most varied in the system, though unhappily some of the most unusual is being savagely removed in favour of standard dull gear, Another mark of the L & L is the number of staircase locks, ranging from the famous "Bingley Five-Rise" through others of three and two.

Starting from the docks in Liverpool, this is the least used length, but much effort has gone into making it more pleasant – apart from the vandals. The Rufford Branch runs down to the R. Douglas with 7 locks in 7 miles. Then the main line goes on to Wigan, where most people join it on its Leigh Branch from the south. There

is a modern, rather garish "development" there now which has certainly enhanced the canal, with its 23 locks upwards in just over 2 miles.

On now through towns such as Blackburn, Burnley (with a high embankment), and Nelson, and to the highest level with a tunnel and vast views. Skipton is delightful, as the canal enters Yorkshire. There is a rest from locks, but not from the 25 swing bridges on the 24-mile level length. Past Keighley come the Bingley staircases (the famous five and another of three), then others on the way near Bradford and into Leeds. Here this marvellous canal joins the Aire & Calder.

(From Liverpool to Leeds. 141 miles with branches, 104 broad locks, over 50 swing bridges).

33 Llangollen Canal

Probably the most popular canal, I'm afraid it isn't my favourite. The very popularity makes it almost too busy to enjoy at times, with boats appearing round sudden bends. Although it is happily rural, the Welsh "flavour" of it – i.e., mountains – doesn't really appear until the far end. It's quite shallow in places, too, and narrow, especially at the top end. There are some attractive lifting-bridges, and of course the spectacular Pontcysyllte and Chirk aqueducts as well as three tunnels. There is a three-lock staircase near Whitchurch. There are few settlements near, though you can go on a short branch into Ellesmere. You also pass the Montgomery Canal, now being restored, before the last striking length via Pontcysyllte to Llangollen itself, high above the R. Dee. There is an interesting small museum here, and horse-drawn trip-boats. Cruise there out of season for the most relaxed journey.

(Hurleston to Llantisilio. 46 miles, 21 narrow locks).

34 Manchester Ship Canal

This canal is a mighty late-comer, built in 1894 but still staggering for its time. The whole canal is the "Port of Manchester". It isn't normally used by pleasure boats, though rather more seem to be having a go since ships declined on it. But you have to get special permission to use it, with regulations about having a survey of your boat, ropes required, and so on.

The surroundings of the upper reaches, which are linked to the Bridgewater Canal by Hulme lock, are being developed for tourists. The canal runs through some very large locks and under huge swing-bridges past Warrington, then alongside the Mersey. The Shropshire Union Canal joins it at Ellesmere Port by the Boat Museum, and it finally runs into the Mersey at Eastham. You'll be lucky to see a ship these days, as most modern ones are too big for the canal.

(From Manchester to Eastham. 36 miles, 5 vast locks).

35 R. Medway

This isn't connected to the system except via the tidal Thames, which requires skill for pleasure boats to use. But there are many boats on it, and it is a very enjoyable river running through the orchards, hop-fields, oast-houses and occasional pubs of Kent. Maidstone and Tonbridge are the main towns on it, and below Allington sea-lock you can visit Rochester if you have tidal skills and knowledge.

(From Allington to Tonbridge non-tidal. 18 miles, 9 extra-broad locks).

36 Macclesfield Canal

A rather late-dug canal, this has fine scenery, and all its locks except the entrance one are in one group of twelve. It is now part of the so-called "Cheshire Ring", which contrasts Manchester cruising with totally-rural Cheshire. Leaving the Trent & Mersey through a short-drop stop-lock, it is soon offering views over the plains one way, and up into the moors the other. It passes above Congleton and then up the 12 Bosley locks, which – unusually – have a pair of gates at each end. The view back is immense. Above Macclesfield then (it's "above" most places), and you cruise on through Bollington to the terminus at Marple, joining the Peak Forest Canal there. There used to be several swing-bridges, but most have gone. What will remain in your memory, though, will be the "turnover" bridges, delightfully snail-shaped ones which enabled the horse to cross over without having its towline untied.

(From Hardingswood – Trent & Mersey – to Marple. 27 miles, 13 narrow locks).

37 Middle Level Navigations

I hardly know where to start here, since this is a whole network to itself comparable with the Birmingham Canal Navigations, but absolutely different. Strictly called "Drains", these waterways are in fact all over a large fenland area, and serve to keep it drained by means of pumping stations which once were a sea of windmills. Now they are not easy to spot, worked electrically after a diesel period.

You can enter these waters from the R Nene at Peterborough, or from the Great Ouse, and cruise below sea-level behind high banks. Climb on your roof to see vast flat distances, with few signs of life across the sugar-beet fields. Farms are dotted here and there. There is only the odd village, especially Dutch-like Upwell and Outwell, and the interesting little town of March. Most people merely cross the Drains to get from the Nene to the Ouse, but this is a pity. The network has much weed in summer, so go in spring. In this way you'll miss the anglers, too, who are thick on the banks during the season.

(A network of waterways east of Peterborough. 92 miles with Old Bedford River, 7 locks, between "broad" and "narrow").

38 Monmouthshire & Brecon Canal

Separate from the network, this re-named canal is really the Brecon & Abergavenny, plus only a bit of the old Monmouthshire. To restore the rest would be a massive task because of the number of locks. The present length is a wonderful leisurely trip, halfway up hillsides.

The Monmouthshire & Brecon Canal is in delightful countryside, part-way up the mountains over the Usk valley.

Thus there's the valley of the Usk below with views across it, and mountains up above you on the other side of the canal. There are several hire-bases if you don't trail a boat, and spring is the time to go, when flowers are lush around you and the greens are unbelievable. There are only six locks, five of them together, and some simple lift-bridges. Abergavenny is well away, in fact, and there are few villages near.

(From near Brecon to the outskirts of Pontypool. 33 miles, 6 locks – between "broad" and "narrow").

39 Montgomery Canal
Like the Huddersfield and the Rochdale, this is a massive restoration under way, after Acts of Parliament, etc., to authorise it. It has been one of the busiest places for volunteer work by the Waterway Recovery Group and the Shropshire Union Canal Society. It will add some wonderful cruising country to the busy Llangollen, and already lengths are open, with trip boats running. There are many problems, such as lowered bridges, but there is much enthusiasm and support.

(From Welshpool to the Llangollen Canal at Frankton, 22 miles, 16 locks. 11 further miles and 7 locks above Welshpool to Newtown Pump House).

40 R. Nene
A complete contrast to the above, this is a river running from Northampton towards the Fens, through Peterborough and later out to the Wash. You can join it by leaving the main Grand

Union and going down 17 narrow locks of its Northampton Arm. Its striking features are the great guillotine gates at the bottom of almost all the locks. Some have been changed, and two massive curved ones have gone, leaving only one example of this unusual type. Opinions vary about the work of using the locks – some people swear, some enthuse. The point is that to raise a guillotine calls for between 80 and 156 turns of a large handle. Then you repeat this to lower it. But at least the turns are reasonably smooth. The ordinary gates at the top ends used to have Grand Union type paddle-gear, but now have worm-drive with far more turns to work them.

You must have an Anglian Water licence for this river, with a small key provided to unlock the guillotine gate mechanism. The river can be tricky if there's been much rain, as there are some low bridges and the extra water may cause a fast current. The scenery is happily varied, from big towns to such delights as Oundle, which the river encloses with a horseshoe. Many little villages stand back from the meadows, and offer good walks. There are lots of Canada geese as well as other waterbirds, and fields full of cattle and sheep. Peterborough has good moorings along a park by the cathedral. Five miles further on the tidal length starts, but before that there is the link with the Middle Level.

(From Northampton to below Peterborough. 66 non-tidal miles, 37 locks – almost "broad").

41 Oxford Canal

One of the most popular canals, and thus busy in summer, with maybe queues for locks. It is in two very different parts – the northern length which leaves the Coventry Canal near Coventry and runs to the Grand Union at Braunston; and the Southern Oxford, combined with the Grand Union for five miles, then meandering lonely for most of its way to Oxford. The northern part had many of its curves cut out at one time to help with trade, losing 14 miles. You can identify these straight lengths clearly. The southern part still has Brindley's wanderings to follow the contours, with one huge hairpin especially.

Lift-up bridges are a hallmark, though many are not used much now, and stay up. there are enjoyable villages and pubs near, but the part through Banbury has never been very well supported by the local council, and mooring is not easy. South of Banbury the locks have a single bottom gate instead of the more common pair. The canal creeps into Oxford, where it is hardly known. But with an effort you can reach this famous city, and there is a link with the Thames.

(From the Coventry Canal at Hawkesbury to Oxford and the Thames. 77 miles, 42 narrow locks).

42 Peak Forest Canal

There were once hopes of linking this canal across the Peak District with the Cromford Canal in the east, but this was eventually done by a now-closed railway. The upper part of the Peak Forest is from Whaley Bridge and Bugsworth Basin to Marple, with staggering views across the valley. Then down its 16 deep locks in only a mile and a fine aqueduct (all restored not so long ago) it falls towards Manchester. This part is more built-up, and there are two short tunnels. It joins the restored Ashton Canal at Dukinfield, where there is also a junction with the Huddersfield Narrow Canal, now being restored over the Pennines. The length from Marple to the Ashton is now part of the currently-named "Cheshire Ring".

(From Whaley Bridge and Bugsworth to the Ashton canal. 15 miles, 16 narrow locks).

43 Pocklington Canal

Although far from the main system, this can in fact be reached via the tidal Yorkshire Ouse and the Derwent. It is a short canal, only partly in use yet, but much-loved by a keen restoration group.

It is very rural, with nature reserves around, and it has some restored swing-bridges. About half is back in use, and hopefully it will again reach within a mile of Pocklington, which is as far as it ever went.

(From the R, Derwent to near Pocklington. 9½ miles, 9 broad locks, not all usable yet).

44 Ripon Canal

Short and sharp, and not fully restored yet, this is really at the end of the system, the northernmost point you can reach without leaving the waterway network. It belongs to the British Waterways Board, and has to be visited via the Yorkshire Ouse and the R. Ure. There is a flourishing Boat Club on the canal and an active restoration society working to reach the basin below the cathedral once more.

(1¼ miles, 3 broad locks, not fully in use yet).

45 Rochdale Canal

One of the two spectacular trans-Pennine canals being restored, this will link the Calder & Hebble with Manchester. Until recently there was only a short (and expensive to cruise) length open in Manchester – apart from a few yards at Sowerby Bridge at the other end. But with a vigorous Canal Society, lengths along the way are coming back into use as fast as you can say Jack Robinson, with already a hire boat and boat trips on them. The locks are broad, and there isn't the problem of a massive tunnel as on the Huddersfield Narrow Canal. Some striking scenery awaits us when we can cruise the whole length. The nine-lock part in Manchester is a section of the "Cheshire Ring", and asks for a hefty fee to use, since the canal is privately owned.

(From Sowerby Bridge – Calder & Hebble – to the Bridgewater Canal in Manchester. 32 miles, 92 broad locks).

Early explorer on the highest level of the Rochdale Canal over the Pennines, once-derelict but now being rapidly restored.

46 Selby Canal – see Aire and Calder

47 River Severn

A comparatively massive river, to be treated with caution after rain, which can make it high and fast. But boats often use it normally in summer, since it has links with two narrow canals, the R. Avon, and the Gloucester-Sharpness Ship Canal. There are five huge locks, with a smaller one into Gloucester docks. They have traffic lights and keepers in high cabins, not to mention mirrors for them to see boats below them out to sight. Stourport is still a fascinating canal town; then you have Worcester, Upton-on-Severn and Tewkesbury as well as Gloucester to call at. Moorings often aren't easy, and of course the river is deep and flowing. There are plans to extend the navigation northwards.

(Navigable now from above Stourport to Gloucester – and of course below Sharpness for suitable boats with proper knowledge. 42 miles, 5 large locks).

48 Sheffield & S. Yorkshire Navigation (and New Junction).

There's a lively bit of recent history to this one. After a long campaign, the government allowed money for most of it to be enlarged to take 700-ton barges to Mexborough, and 400-tonners to Rotherham. The rest to Sheffield was not enlarged, so still has ordinary broad locks, heavy to work. The part from the Trent to Bramwith wasn't included either, since the New Junction connects there with the Aire & Calder. it was hoped that a busy trade would build up with the North Sea ports, but this has been slow to happen. Meantime, the canal is a delight to cruise, since all the locks to Rotherham are worked by keepers in their pleasant cabins. You enter it officially at Keadby on the Trent (take good advice on getting there), and the first – unenlarged – part has swing-bridges worked for you (including a fascinating railway one) and three locks. Check on the times when they are usable, especially at weekends. Call at Thorne for good shops.

Climbing up the final 11 locks to Sheffield — heavy work but worth the effort.

From Bramwith you are on the improved part, cruising grandly, along with much rurality and much industry also. You pass Doncaster, Mexborough and then enter Rotherham with the non-improved locks ahead (supermarket by the one at Rotherham). It's heavy going up the 15 locks to Sheffield, but you're climbing to wonderful industrial views. The basin at Sheffield is the subject of development at long last, but how appropriate will it be?

(From the Trent to Sheffield. 43 miles, 28 locks – mostly large. New Junction Canal link with Aire & Calder, 5 miles, 1 lock).

49 Shropshire Union Canal

A very popular waterway, which is shallow in parts, running from the Black Country rurally to Chester and Ellesmere Port, with a branch to Middlewich and the Trent & Mersey. Indeed, the Llangollen is strictly SU also, as well as the Montgomery which is under restoration. You have delightful countryside most of the way, with little villages near, as well as towns at Market Drayton and Nantwich, then of course the wonders of Chester, not to mention, at the end, the remarkable Museum at Ellesmere Port and the Manchester Ship canal beyond.

There's a lovely flight of locks at Audlem and a little town, and some great cuttings and embankments, since much of this canal was cut more direct by Telford than Brindley's earlier ones. The northern locks near Chester are broad, with a staircase of three in Chester itself, and a link with the R. Dee, which is not really suitable for normal cruising.

(From Wolverhampton, Staffs and Worcs Canal, to Ellesmere Port, Manchester Ship Canal, with a branch to the Trent & Mersey. Main line 66 miles, 46 locks, mostly narrow. Middlewich branch 10 miles, 4 narrow locks. Branch to R. Dee short with 3 locks).

50 River Stort – see Rivers Lee & Stort

51 Staffordshire & Worcestershire Canal

Another long and varied one, running from the Trent & Mersey in the north to the Severn in the south. It skirts the Black Country about half-way, but the rest is pretty rural apart from Kidderminster. Lots of locks, nicely spread out except for an odd "semi-staircase" of three at Bratch. There's a wide "lake" near the northern end, and some rolling scenery all the way. Nice villages around, and good pubs alongside. Notice the bridge name-plates, many restored by the busy canal society. The Shropshire Union comes in quite near the link with the BCN, then you are soon into sandstone country. There is even a little cave by one lock. There's a southern link, via the Stourbridge, with the BCN, then very wooded hills by Kinver and on to Kidderminster. Stourport, at the Severn junction, is a true canal town with basins, boats, buildings, and alternative broad and narrow locks to the river.

(From the Severn at Stourport to Great Haywood, Trent & Mersey. 46 miles, 45 locks – all narrow except for 2 broad ones available at Stourport basin to the river).

52 Stourbridge Canal

A short canal, but with many locks, linking the Staffs and Worcs to the Birmingham Canal Navigations. From country to industry, with fine backward views as you climb.

(From Stourton to Black Delph, BCN, with short branch into Stourbridge. 7 miles, 20 narrow locks).

53 Stratford-upon-Avon Canal

It is an odd story because the southern part was completely derelict until restored in 1964, from when it was run by the National Trust. Now it has returned to British Waterways whose northern length was never closed.

The southern restoration was historically famous, with David Hutchings, voluntary labour and voluntary money bringing it back to life.

The whole canal runs from the Black Country to the Avon outside the Shakespeare Theatre. It starts with an unusual – now always open – guillotine "stop-lock", then is industrial at first, and level for ten miles before starting its fall through some very close locks, easy to work.

There's a short link with the Grand Union before the southern section, which has unique "barrel-roofed" lock cottages for some way, and some unusual aqueducts. Another set of close locks brings you to Stratford, creeping past the back of the town to the glorious basin and the lock to the Avon.

(From Kings Norton to Stratford. 25 miles, 55 locks – all but the river lock are narrow).

54 River Thames

It is impossible, of course, to describe this great river decently in brief. In every way it's a law to itself, including the very different types of boats to be seen on it. Most are big and gleaming, though canal boats can be seen too. Locks are manned by smart keepers, and are often a blaze of flowers. There's both history and bustle all along the way, though the river above Oxford is quieter and more like the remoteness of many canals. Below Oxford the Thames is often too much for quieter inland boaters, with its mass of big cruisers, its even bigger passenger boats, and the overwhelming tourist atmosphere of such places as Windsor and Marlow, as well as Oxford itself of course. But if you'll accept the queues at locks in summer, you can have a leisurely cruise with no work to do. Be ready to pay to moor, and even have difficulty finding anywhere. Nor will you find food and drink very cheap.

There are of course very many places to hire boats, from rowing boats to the massive cruisers. It is the non-tidal part above Teddington that is normally used by inland pleasure boats, but you can go down-river and reach the Grand Union Canal at Brentford, then go on spectacularly through London to join the R. Lee as well as the Regents Canal at Limehouse. But this section calls for tidal knowledge and advice.

(Non-tidal from Teddington to Lechlade is 124 miles, 44 locks – wider than "broad" below Oxford. From Teddington to Grand Union at Brentford 5 miles, and a further 16 to Limehouse).

The sophisticated Thames has well-kept locks with lock-keepers, and patrol boats such as this one.

55 River Trent

Another mighty river, but very different from the Thames. A bit of it in fact flows through the Trent and Mersey Canal at Alrewas, but the main navigable part starts at Shardlow between Nottingham and Derby, the end of the Trent and Mersey. You have to leave the river to pass through Nottingham by canal, and join it again down to the Humber (though this lowest part is not recommended without advice). There are well-spaced vast locks, now all mechanized, though the recently-converted ones must still be worked by hand (and "work" is the word!) after normal hours. The rest are manned quite late by keepers.

The canal in Nottingham gives access to shops, including a supermarket alongside now, but the river then is wide and often empty. Newark is a good calling-place, and not far below it you reach the tidal part, where advice is vital. There's a link with the Fossdyke and Witham, with Lincoln worth the trip, and another beyond Gainsborough with the lovely Chesterfield Canal. The last junction for most inland boats is at Keadby with the Sheffield & S. Yorkshire Navigation. The lock here is tricky to enter, and you must let the lock-keeper know in advance that you are coming. There are 9 more miles to the Humber, but you need expert knowledge to use this part. A mighty river, as I say, but with care and sense an enjoyable and certainly uncrowded cruise.

(Navigable from Shardlow to the Humber. 95 miles, 13 locks – broad in canal sections, much bigger on main river).

56 Trent and Mersey Canal

This is a major canal, though it no longer joins the Mersey. It is a historic one, once called the "Grand Trunk" across England. It has enormous variety, from a great and distinguished tunnel to a "heartbreak hill" of locks, from truly rural views to industrial towns ranging from brewing to pottery. There are little villages, pubs a-plenty, broad locks at one end, but gentle narrow ones in other places, and junctions with several other waterways. From the Trent and the canal-and-boat-building village of Shardlow you soon reach Burton (and smell the beer!). Then there are easy narrow locks on past the *Swan* at Fradley and by a modern colliery at Rugeley, where the ground keeps sinking. Junctions point towards the Midlands, then you're on your way to the Potteries, birthplace of the canal. The pleasant Caldon Canal leaves here.

Surroundings are built up for some way before famous Harecastle tunnel, one-way so watch the times. You work down Heartbreak Hill now, to Middlewich, with its branch to the Shropshire Union. Soon after this you begin to have views down to the R. Weaver, with Anderton Lift leading to it. There are two crooked and one long straight tunnel along here, before you come to the Bridgewater Canal and points north.

(From Shardlow, R. Trent, to Preston Brook, Bridgewater Canal. 93^1/$_2$ miles, 76 locks, mostly narrow).

57 River Weaver

A very different matter from the narrow canals, since you'll see ships. But it's well worth a visit, deep and wide but not tidal. For very many years the usual entry for inland boats was down the remarkable Anderton Lift from the Trent & Mersey. This has been in trouble, but is being restored as I write. It's an incredible ride, in a tank of water, to the river below. Once down, you may share large locks with ships, but they are kind to us. There's quite a bit of industry, especially at the bottom end, and even a salt-mine alongside further up. Above Winsford you enter a large lake-like area called a "flash", but beware of shallows, and don't try to reach the sides.

(From Winsford to Weston Point, Manchester Ship Canal. 20 miles, 5 large ship locks).

58 River Wey

A tributary of the Thames, this is run by the National Trust. It is a pleasant cruise seemingly past built-up areas, but you often don't see them. It is navigable from Godalming, and flows through Guildford with spread-out locks and lots of woodland. The Basingstoke Canal joins it near Byfleet, and there are many visitable places along the way, such as the Royal Horticultural Society's gardens at Wisley. The river at last passes by Weybridge on an artificial cut to the Thames near Shepperton.

(From Godalming to the Thames. 20 miles, 16 broad locks).

59 Witham Navigable Drains

Not many people have boated on these waterways, which are comparable to the Middle Level in some ways. But they are more remote to reach, and more difficult to move about in. Some of the tail-ends are weedy and derelict, with vanished locks and sluices, but you can enter at Antons Gowt lock from the R. Witham near Boston. Using guillotined Cowbridge lock (ask the Water Authority in advance how to obtain the key) you can moor in the heart of Boston. Otherwise it's a lonely area with not a soul in sight.

(Network near Boston. 80-odd miles of sorts, 2 locks).

60 Worcester and Birmingham Canal

Famous (or notorious, according to your viewpoint) for having 58 locks in 16 miles, some very close. The Tardebigge flight has 30 locks, but in fact there are another 12 quite soon, so that makes 42 – in 5 miles! They all start their drop at Tardebigge after a 14-mile level from the heart of Birmingham.

Apart from the Birmingham end, it's all quite rural, with vast views in places down towards the Severn valley. After the 42 locks you have time to look around, the rest being more spread out. There are several pleasant calling-places and pubs, with a friendly hire base at Alvechurch for example. There are also five tunnels, one being 2,726yds long, and two others over half-a-mile. You come to Worcester at last, with its interesting basin, and two broad locks into the Severn, where the cathedral dominates the view.

(From Birmingham to Worcester. 58 locks, all but two being narrow).

61 River Yorkshire Ouse and River Ure

Northwards from Naburn locks (just below York) this grand river is tideless, and there are plenty of boats glued to the banks. But of course the usual care must be taken as on any non-tidal river. Getting to this area from the rest of the system is a different kettle of fish, and requires some planning. It can be done from the Aire & Calder at Goole, but this is especially dependent on tidal knowledge. The idea is to reach Naburn before the tide starts returning. The best bet is to join the Ouse at Selby, on the advice of the lock-keeper there. Look out at the two big swing-bridges in case a ship appears. After that it's a 14-mile rather featureless slog to Naburn (did you ring the lock-keeper from Selby?).

Naburn Marina will sell you an essential chart, then there's a lovely mooring in York quite near to the Minster. It's quieter above York, but watch out for "clay huts" below Linton lock – humps in the river which you may slide off. The chart marks them. There's a fee for Linton, well-deserved, for its supporters have had a struggle to keep it open. Above Linton the river becomes the Ure at an unnoticed spot where the Little Ouse Gill Beck comes in. The Ure has two locks before the Ripon Canal (q.v.).

(From the Humber – and Trent – at Trent Falls to the Ripon Canal. Tidal from Trent Falls to Naburn, 37 miles. Naburn to Ripon Canal, 31 miles, 4 large locks).

Other waterways

There are of course many separate and tidal rivers where you can use small boats towable by car. Several are on the east coast and rivers such as the Arun and Adur in the south are used by enthusiasts with careful planning. But be very wary of tides with small inland boats.

There are also surprising numbers of true inland waterways in various degrees of restoration. Some are well on their way, others are still optimistic gleams in the eyes of their society or trust.

Here are some where various things are happening, but somebody will up and tell me of others lost in the fields somewhere:

Barnsley (and Dearne & Dove) Canals
Chichester Canal
Derby Canal
Driffield Navigation (with R. Hull and Beverley Beck)

Droitwich Canals
Edinburgh & Glasgow Union Canal
Forth & Clyde Canal
Grand Western Canal
Grantham Canal
Hereford & Gloucester Canal
Ipswich & Stowmarket Navigation (R.Gipping)
Manchester, Bolton & Bury Canal
Sankey Brook (St. Helens Canal)
Sleaford Navigation
Tennant & Neath Canals
Wendover Arm (Grand Union)
Wey & Arun Canal Trust
Wilts & Berks Canal
If you feel like it, join one of their societies and help them along!

Not a normal "inland waterway", but tidal here — the R. Arun. It may well, however, become again part of a restored link with the R. Wey if the Wey & Arun Canal Trust has its way.

13

FINDING YOUR WAY AROUND

Where am I?

I once met a man who had just bought a small boat on the Shropshire Union Canal and was setting off in it. He asked me the way to Wales and it appeared that he hadn't a map or guide of any kind with him. He was just ambling off along the canals, and had a rough idea that Wales was somewhere along the way from where he was.

Fair enough, for I can't think of a better way of getting away from the rat-race. Indeed, that way you can get so far away that, like him, you end up without the faintest idea where you are. Most people, though, do at least like to have some notion of which way they're going, roughly where they are at any one time and even where the next pub is. So they need a guide.

I know some boaters who just carry the Landranger Ordnance Survey maps of the area, and get along quite well with them. But these maps do tend to hide the canals in places and the locks are rarely shown clearly or correctly. And of course they give no indication of what cruising along a waterway is really like, what the pubs offer, what to see or buy at the local village, and so on, the things any true "guide" should tell you.

A guide will give you an idea of what the actual navigation is like, how wide the locks are, which way they rise, bridge-heights, where

there are boatyards, what to see locally, and other details about your surroundings that the map can't really reveal. So don't set out blindly, unless that's what you want to do.

The most comprehensive guides are the ones familiarly known as "Nicholsons". The original ones were issued in conjunction with the British Waterways Board, but the latest are in co-operation with Ordnance Survey, and thus include waterways not under the Board. I'll deal with these guides first.

Nicholson/Ordnance Survey Guides

These guides really do tell you everything you'd like to know. They show the waterway down a long page at two inches to the mile (except in the north-east) with its width exaggerated so as to make it very clear. All the nearby villages and towns are shown too. There are clear symbols for locks, aqueducts, tunnels and such features and almost always the lock-depths are shown. You're given bridge-numbers or names too, a helpful clue to where you are.

The guides mark places where you can get water, deal with your toilet, and turn a long boat round, and the text alongside the maps gives much useful information. It tells you, for example, about boatyards and their facilities, pubs and their facilities, and what each village has to offer in the way of shop, phone box, post office, garage, and places worth seeing.

There is also a special text section on the actual navigating of the length on each map-page, partly what you will see, but also any particular point about cruising, such as stiff swing-bridges, lock details, narrow parts, and so on. There are useful telephone numbers, potted cruising hints, and even a section about fishing. All in all, there is a mass of helpful information for anyone setting out on a cruise. Five of these guides will now be listed.

Nicholson/Ordnance Survey Guide to the Waterways: 1 South

This covers all the navigable inland waterways roughly south of Birmingham, apart from the Thames, Broads and Fens, which have separate guides. Thus it deals with the London area, the Kennet & Avon, the isolated Monmouthshire & Brecon, and the network of the Grand Union, Oxford, R. Avon and Stratford Canal, and the Worcester & Birmingham Canal, together with the Severn.

Nicholson/Ordnance Survey Guide to the Waterways: 2 Central

This overlaps the South and North guides a little. It covers the Birmingham Canal Navigations and those canals linked near them, such as the Staffs and Worcs, Shropshire Union, Coventry and Birmingham & Fazeley. It also includes the Ashby, Trent & Mersey, Caldon, R. Weaver, Llangollen, and the Leicester line of the Grand Union.

Nicholson/Ordnance Survey Guide to the Waterways: 3 North

This one gives rise to criticism that some Yorkshire waterways are not covered as thoroughly as the rest. The publishers must feel that not enough pleasure-boats visit these waters, yet that attitude may even help to keep them away if there is no detailed guide. In fact the commercial waterways in Yorkshire are fascinating and not hazardous. Even the big locks are worked for you, and usually have less turbulence than some of the locks elsewhere.

Anyhow, this guide covers waterways north of the ''Central'' one, that is, part of the Trent & Mersey, the Macclesfield, Peak Forest, Ashton and Bridgewater, and the Leeds & Liverpool. To the east it deals with the R. Trent, the Erewash, the Fossdyke and R. Witham, and the Chesterfield. It covers the isolated Lancaster, then in Yorkshire it looks briefly (shame!) at the Calder & Hebble, Aire & Calder, R. Derwent, Pocklington Canal, Yorkshire Ouse, and the Sheffield & S. Yorkshire Navigation.

Nicholson/Ordnance Survey Guide to the Broads and Fens

A much-needed addition to the series, this now offers a useful guide to an area of waterways which had a miscellaneous collection of odd guides, some not very detailed outside the Broads. It has the usual immense amount of detail for the whole Broads area, and an even more welcome lot of information about the R. Nene, the Middle Level, and the Geat Ouse and its tributaries. Boaters who leave the canal system down the Nene will be glad of this one.

Nicholson/Ordnance Survey Guide to the River Thames

The usual conscientious approach is made to this very different waterway, with its sophisticated boats and boatyards and its smart locks and lock-keepers.

Waterways World Guides

These in contrast to the above, and published by the long-established waterway magazine, tackle the need by having a separate booklet for each (or small groups) of the more popular waterways. The booklets are thus less bulky than Nicholsons, and you can use just the ones you need for any one cruise. They are made with ''wiro'' binding to lie flat when opened, and the waterway enthusiasts who produce them, offer you all you'll want to know about each route. The guides show not only the essential boating needs, but railway stations, fish and chips, off-licences, bus stations and banks. There are numbers at each waterway mile along the map (which Nicholson unfortunately doesn't have) and even an indication of where towpath walking is poor. Advertisements may point you helpfully to restaurants, bakers, supermarkets, launderettes, etc., as well as to nearby pubs. There's an interesting historical introduction, a bibliography, a list of local waterway societies (though addresses quickly date), and of course full cruising help and details. Again the 2-miles-to-the-inch route is accompanied by notes of what to see, what to be beware of, and so on.

Titles available are: *Coventry, Ashby and Oxford (N), with Birmingham & Fazeley Canal. Grand Union Canal North (Birmingham to Stoke Bruerne). Grand Union Canal South. Oxford Canal, from Coventry Canal to Oxford. Llangollen Canal (with Mongomery Canal). Shropshire Union Canal. Staffs and Worcs Canal. Trent & Mersey and Caldon Canals. Leicester Section, Grand Union, with R. Soar and Erewash Canal.*

Pearsons Canal Companies

This small set of guides, apart from one, has yet another approach. It looks at popular "Rings", where you can set out in one direction and return from another. I always think this is a bit risky if you are hiring a boat, since timing may go haywire if there are any snags, or queues at locks, and you then may not get back in time. And some "Rings" are just that little bit too extensive for leisurely cruising, whereas on a straight "out-and-back" cruise you can turn round at half-time wherever you've got to. Moreover, the view on the way back is entirely different. However, the "Ring" appeal (some say "mania") is well-established now, and these guides will guide you round six of them. Again compiled by waterway enthusiasts, they show all you want, and have some interesting informative articles too.

Titles available are: *Cheshire Ring; Warwickshire Ring; Four Counties Ring; Avon Ring; Stourport Ring; Black Country Ring; Llangollen and Shropshire Union Canals.*

Other Guides

There is a whole collection of guides to individual waterways, often produced by the local society, varying greatly from duplicated sheets to well-printed booklets. They come and go, unfortunately, as some run out of print, or an enthusiastic compiler moves on to other interests. Thus it would be unhelpful for me to list all those available as I write, since some certainly won't be there when you read this.

Luckily, most such guides are obtainable from the Inland Waterways Association sales department, or from *Waterways World*, whose addresses I give in the next chapter. They issue regular lists of those they have. I'll just mention a few of the outstanding ones, which are pretty likely to continue for some time, and if you are interested in others, the IWA and *WW* lists will help you.

The Kennet & Avon Canal, by Niall Allsop. A real book, with full charts of this restored waterway, and much enjoyable reading material also.

London's Waterway Guide, by Chris Cove-Smith. Another fat book, with incredible detail about London's canals and the Thames, even listing a wealth of information such as nearby bus-stops.

West Yorkshire Waterway Guide. Drawn up by the Calder Navigation Society, this helps very much to fill the gap left by Nicholson's thin coverage of these waters.

Northeast Waterways, by Derek Bowskill. Also covering Yorkshire waterways and others nearby, this is not only a guide with maps and lock-plans, but also a very entertaining personal account of cruising there.

Birmingham Canal Navigations. Produced by the Birmingham Branch of the Inland Waterways Association, this gives minute details of this network of waterways, with surrounding streets and all the shops, fish-and-chip parlours, and everything else you need.

Among the smaller guides which may continue to be available are the following: *The Upper Avon Navigation, Gateway to the*

[Lower] Avon, Chesterfield Canal Guide, Coventry's Waterway, The Derwent [Yorkshire] Guide, Basingstoke Canal, Bridgewater Canal, Stratford-upon-Avon Canal, Worcester & Birmingham Canal, Erewash Canal.

Maps

Although the above guides include strip-maps of waterways, it is also helpful to have more comprehensive actual maps of the network or parts of it. Here follow some which are especially useful:

Imray's Inland Waterways of England & Wales. Large wall-map of the whole system, with quite a bit of symbolic and text information about the different waterways.

Nicholson/Ordnance Survey Inland Waterways of Great Britain. Similar to the above, with Scotland too, and boatyards, etc., shown by symbols. Main roads are also clearly shown.

Lockmaster Maps, by Douglas Smith. These are delightful maps which can almost be used as guides also. Detail is in rather small print, but each map covers a waterway area (e.g., Black Country Canals, Cheshire Ring, Fenland Waters), and they show all the kinds of things boaters like to know about. Distances are given at regular intervals, and the helpful canal bridge-numbers. Measuring about 20 inches by 18 inches, or rather less, there are now 25 of them, covering practically the whole waterway system.

Imray Maps. This firm publishes maps of individual waterways. The following are likely to be available: Cam and Lower Great Ouse, Upper Great Ouse, R. Medway, R. Nene, Middle Level, R. Wey, R. Thames (Teddington to Southend), Northeast Waterways.

Other Maps

Three other useful maps can be obtained from the IWA or Waterways World. They are: Kennet & Avon: East, Kennet & Avon: West and Stanford's Map of the R. Thames: Lechlade to Teddington.

14

BITS AND PIECES

I'll end now with various pieces of information which may or may not be helpful, according to the amount of enthusiasm that you've developed. Among the addresses are some that deal with boat licences.

Addresses

The Inland Waterways Association, 114 Regents Park Road, London NW1 8UQ. This is the body of dedicated volunteers without which there would be very few waterways for you to cruise on. Since 1946 it has battled with authority, educated the public and governments, and completely turned round public and official attitudes. Thus our present magnificent water network is very much composed of what were once rapidly becoming stinking or dry ditches. The 'IWA' has grown into a vast national body now, with 32 branches around the country and you are welcome to join by writing to the above address. **N.B.,** this is also the address for the *Sales Department* from which you can get most of the guides and maps mentioned in the previous chapter and the books mentioned later in this one.

Waterway Societies and Trusts. The IWA has given rise to 50-plus separate local bodies dedicated to "their" waterway. Some are less active than others and the busiest ones are those still working for the restoration of various canals and river navigations. Examples of these

are the Huddersfield Canal Society and the Rochdale Canal Society. The greatest such organisation is the Kennet & Avon Canal Trust, whose massive work is almost done. The Surrey & Hampshire Canal Society has also virtually completed a long slog with the Basingstoke Canal. Secretaries come and go, so it would be unwise to list addresses. Ask along any waterway, or the IWA will have details of any society which interests you.

British Waterways (HQ) Melbury House, Melbury Terrace, London NW1 6JX. (For licences) *Willow Grange, Church Road, Watford, Herts WD1 3QA.* This body runs almost all canals and some rivers. It has regions around the country and sections within each. There you find maintenance yards and offices. Each region has a special leisure department. As you cruise the canals you constantly meet BW people and boats working on the waterway, dredging and repairing, dealing with locks, tunnels, bridges, and so on.

There are many other authorities controlling odd rivers and the occasional canal, most quite small. But most have a licensing system. The most important deal with the Thames, the Broads and a number of eastern rivers.

The Thames from London to Lechlade is under the *Thames Water Authority, Nugent House, Vastern Rd., Reading RG1 8DB.* For the Broads, write to *The Port & Haven Commiss-*

ioners, 21 South Quay, Great Yarmouth, Norfolk NR30 2RE. For the R. Nene and other Anglian Water rivers, write to *Anglian Water, North Street, Oundle, Peterborough PE8 4AS* (Nene authority), or *Anglian Water, Great Ouse House, Clarendon Rd., Cambridge CB2 2BL* (Ouse and tributaries).

The Anglian licence can also be obtained from the shop by Buckby top lock, and the boatyard at Blisworth, on the Grand Union Canal, very helpful when approaching the Nene from the canal system. Make enquiries, though, about the "free" Nene cruising you can have with a British Waterways licence.

One smaller authority is of especial interest to users of the main waterway network. To travel the popular "Cheshire Ring" you cruise for 1¼ miles (and 9 locks) on the Rochdale Canal, which was never nationalized. So you pay for this, to the *Rochdale Canal Co., 75 Dale St., Manchester, M1 2HG* – and a fine penny, too.

Incidentally, you also use the Bridgewater Canal on this trip, and this too is a separate body. But luckily a British Waterways licence is accepted for through journeys. Boats which stay on the Bridgewater need a licence.

Other interesting waterways calling for extra licences are the Upper Avon and Lower Avon (Warwickshire), the R. Wey, the Basingstoke Canal, the Manchester Ship Canal and to pass Naburn Lock and Linton Lock on the Yorkshire Ouse. The Middle Level, incidentally, demands no licence. The most helpful source of authority addresses for all waterways is the unrivalled "bible" – "Edwards" (see Books later).

Magazines

Waterways World, Kottingham House, Dale St., Burton-on-Trent DE14 3TD. The oldest-established colourful monthly magazine for inland waterway enthusiasts. Obtainable of course from newsagents or by subscription. It also has a Book Service, offering guides, maps, etc., as already listed, and some other books.

Canal & Riverboat, 9 West St., Epsom, Surrey, KT18 7RL. Similar colourful monthly also for inland waterways, and obtainable from newsagents or by subscription.

Inland waterways are also mentioned at times in some general boating magazines.

Museums

Inland waterways figure as part of many museums, large and small, all over the country. But there are also some devoted entirely to the subject. The main ones are:

National Waterways Museum, Gloucester Docks. This is the newest one, opened by British Waterways in an old warehouse in these fascinating docks. It shows the whole picture of our inland waterways. A "must" for anyone with the least interest in the use and value of canals and rivers.

The Boat Museum, Ellesmere Port, Cheshire.
Far from having just boats, this is a big complex, but with an especially remarkable floating collection of inland boats of all shapes and sizes. It is where the Shropshire Union joins the Manchester Ship Canal, and there are also various special departments, with particular emphasis on educational displays.

Waterways Museum, Stoke Bruerne, Towcester, Northants.
This was the original BW museum, on the Grand Union near the huge Blisworth tunnel and by a lock. It still houses, on various floors of an old canal building, a great variety of things to see and school parties are well catered for.

The Black Country Museum, Tipton Rd., Dudley, W. Midlands.
On a BCN canal, and next to the entrance to Dudley Tunnel, this tackles not only the intriguing history of midland canals, but also past life in many other Black Country spheres. There are old houses, shops, pub, trams, etc., as well as old canal boats and workshops. Trips into one end of Dudley tunnel run from there.

The Canal Exhibition Centre, Llangollen.
Near the terminus of the Llangollen Canal, this

The Waterways Museum is at the far end of these cottages at Stoke Bruerne — still the most intriguing canal museum in many ways, with locks, a pub, and a very long tunnel nearby also.

small museum has a valuable taped commentary and many small exhibits about our canal system.

Ironbridge Gorge Museum, Telford, Shropshire.

There is some inland waterway material here, connected with the Severn and the local canals which used to join it in that area.

Books

There are many waterway books old and new (though not as many as there are for railways). Most of those in print can be obtained from the IWA Sales Dept (address earlier) and a few from Waterways World Book Service (also address earlier). Many, too, especially those out of print, can be bought from *M & M Baldwin*,

24 High St., Cleobury Mortimer, Kidderminster, Worcs DY14 8BY, or from *Shepperton Swan, the Clock House, Upper Halliford, Shepperton, Middlesex.*

Ask the IWA for the latest list of current ones, among which you may like to try the following: *Inland Waterways of Great Britain,* by L.A. Edwards (Imray Laurie). The aforementioned "bible", absolutely essential for every detail about inland waterways, even down to the language used on them (well, the respectable bits). Lengths, dates, locks, boats, maps, authorities, measurements, history, pictures – a massive tome, but it has the lot.

Charles Hadfield's books (David & Charles).

Many "bibles" of a different sort, since these hold the full and detailed history of our waterway system. There is a whole collection of them, essential to the serious student, covering different areas. They get right down even to the minutes of important companies and there's nothing to rival them in the history field. Charles Hadfield has also written a mighty book called *World Canals.*

Observers Canals, by John Gagg (Warne). A pocket attempt to describe canals in every way, looking briefly at the locks, boats, tunnels, aqueducts, and all other aspects of them, with summaries of each navigable one, and lots of helpful information. The opposite in size to "Edwards".

Shell Book of Inland Waterways, by Hugh McKnight (David & Charles). A large book, not so statistical as "Edwards", but containing very readable details about our waterways. There are full sections on the usual locks, tunnels, etc., and also about wildlife, some waterway towns and villages, commercial carrying, maintenance, and so on. The Gazetteer (over half the book) gives a blow-by-blow description of each route, with some vivid word-pictures of much on the way.

Other books you might like to try, are: *Lost Canals and Waterways of Britain,* by Ronald Russell (David & Charles). Over 100 waterways

Some of the many canal books available.

now largely gone. *Narrow Boat Painting,* by A. J. Lewery (David & Charles). The historic decoration of the old narrow boats. *Narrow Boat,* by L. T. C. Rolt (Eyre Methuen). Originally published in 1944, this book sparked off the whole restoration movement. *Exploring England by Canal,* by David Owen (David & Charles). Pleasant account of actual cruises by the widely-travelled author.

Rallies

Among the greater mystiques dear to canal enthusiasts, especially the longer-established ones, are events known as "Rallies". Some are local and small, others national and large. To the outsider they seem to hold many mysteries, but a few minutes at one bring nothing but interest and enjoyment.

These rallies originated in the early days of struggles to restore canals. Grim pioneers battled their way through weed and rubbish to some remote spot where a canal was threatened with closure. These rallies were a powerful

Part of a small rally at Whaley Bridge. The large "National" — held in different places each year — attracts well over 500 boats.

weapon in this fight and the first national gatherings have gone down in history for their impact on opinion. Lately much of this aim has been lost, as the attendance of more and more boats called for bigger and bigger venues. Thus the huge IWA National Rally can hardly be held in a threatened remote spot these days, though there are still some left. And it is fair to say that many boaters attending go more for sheer pleasure than out of any grim campaigning motive. But smaller "campaigning rallies" have now been developed to keep up the old aims. Many canal societies and trusts hold their own small rallies.

The "National", however, now seeks more to raise money for the IWA than to restore a particular waterway. It also shows the public, who roll up in thousands, what canals and rivers are about, and what has been done in the last half century to bring them back into valuable use for commerce and pleasure.

At rallies, therefore, boats gather from far and wide – vast numbers of them at the "National", which is held in a different place each year. In addition to the boats, nowadays, you'll find such things as steam fair engines, Morris dancers, parachutists, performing dogs, hot dogs, candy-floss and dozens of stalls selling all sorts of things, some nothing to do with canals. The boats are the centre of attraction, all the same. To those on them, from the far corners of the waterway system, they provide a grand get-together where you meet all your old pals whom you haven't seen for years (well, since last year's National). To others they reveal a dazzling variety of personal choice, from home-built paddle-wheel-driven boats to magnificent micro-waved palaces, from converted (or unconverted) commercial narrow boats to steam-driven boats. Visitors walk endlessly along the towpath, maybe for miles, to inspect the galaxy of floating variety.

It is impossible to describe this variety and the colour of such gatherings, so you must go to one, small or large. Some stalwarts feel that the big ones have lost their way and are now mere social and money-raising activities. More seriously, the arrival and departure of such large numbers of boats cause problems to others just wanting to cruise peacefully in the area. However, a rally is a colourful event, attracting more and more of the general public and you'll find many, large and small, each summer. The waterway magazines list them well in advance.

Hiring a boat

Now, where do you find a boat to hire? They are in fact available at a remarkable number of places all over the waterway system. The Nicholson/Ordnance Survey Map mentioned in

Boatyard, fuel pump, chandlery and hire boats at Brinklow on the northern Oxford Canal.

Chapter 13 has a boat symbol to show how widespread they are. Many hire-firms advertise in the waterway magazines, and many also are represented by three main agencies. These provide colourful brochures, giving full details and plans of all boats, their location, and their prices at different times of the year. They also give a good description of what it is like to cruise and the flavour of different areas. The agencies are: *Hoseasons, Sunway House, Lowestoft, Suffolk NR32 3LT; Blakes Holidays, Wroxham, Norfolk NR12 8DH; Boat Enquiries, 43 Botley Rd., Oxford OX2 0PT.*

There are of course many other boatyards which operate without using an agency. The best way to discover all of them is to buy an annual book available from the IWA Sales Dept around December/January. This is the *Inland Waterways Guide* (for the current year), and it lists over 400 firms, including those in Scotland and Ireland. It also gives much brief information about cruising and waterways.

Firms will gladly send you their own brochure. Some even have open days early in the year so that you can have a look at what they offer. It really is a good idea to try and see a hire boat if you are a complete novice contemplating having a go, even if it means asking politely when you see one moving through a lock or moored by a towpath. Its users can always say no.

Buying a Boat

Almost all those who now own boats on inland waterways started off by hiring boats and seeing what they thought about it all. Then – like us – they maybe bought a small 25ft glass fibre cruiser, perhaps second-hand and went on from there over the years. A boat is far less harassing and more roomy than a caravan even though you can't usually keep it at home. So if you feel inclined to buy one, how do you set about it?

The immediate first step is to buy *Waterways World* and *Canal & Riverboat.* These magazines not only have many ad-vertisements for hire-firms, as I mentioned earlier, but also many others of boats for sale, new or second-hand. They include firms who will build you a boat to your own needs, boatyards where you can see boats for sale, brokers who deal with boats moored in different places and owners selling privately. You can have a look round boatyards or fix up visits, but I would say one thing strongly if you are looking at second-hand boats: Have a survey made.

You would do this with a house if it is of any age and a boat can get knocked about far more than a house. Moreover, the part under the water can be in all sorts of states. So a surveyor's fee is money well spent. Again, the two magazines mentioned advertise a selection of qualified surveyors who will do this for you. Don't rush at it, though. Work out what you want in a boat, imagine what it will be like using it, make sure you will have somewhere to moor it (boatyards selling boats may offer moorings later) and above all, make sure you know what it will cost to licence, insure and run. Good luck to you.

Some waterway words

It might amuse you if I end with some of the waterway words that have come down to us, often from way back in the old canal days. Some I have already mentioned, especially in connection with lock-working, but others may also interest you. So here's a selection, but (as always) "Edwards" has a fuller list.

Animals. The old boatmen used this word for donkeys which sometimes towed boats, two equalling one horse.
Balance beam. The long arm, usually wooden, sticking out over land to help you open and close a lock-gate.
Barge. This means a boat around 14ft wide and more and thus is wrong when applied to a 7ft wide narrow boat.
Bollard. Wooden or metal stubby posts for tying your boat to. Often seen at locks (but

"Gongoozlers".

don't tie up if you're "going down"!).
Bow hauling. Pulling a boat along by hand, using a long rope.
Bridge hole. The space through a canal bridge.
Butty. A boat without an engine, towed by a boat with an engine (the "motor").
Cratch. An upright, roughly triangular-shaped board near the front of a commercial narrow boat, used to support the front end of the centre

planks which held up the covers. Many pleasure-boaters like to have a cratch, too, beautifully decorated.
Cut. The old name for a canal, for obvious reasons.
Fenders. Used on boats to protect them (and lock-gates, canal edges, etc.) from collision, rubbing, and so on. May be made of rope (especially at the front and back of boats), but most side-fenders are now of white plastic – a strange colour-choice when they will certainly pick up not only dirt but oil and tar.
Gang-plank. A plank which many boaters carry in order to be able to get ashore if they can't get the boat right to the bank when they moor.
Gongoozler. The most amusing waterway word. As I mentioned earlier, it has been defined some long time ago as "an idle and inquisitive person who stands staring for long periods at anything out of the common". Nowadays boaters refer – affectionately – to all idle spectators as gongoozlers. On a fine day you meet many.
Handspike. The length of wood still used to open some paddle-gear on the Calder & Hebble Navigation.
Keb. A long-handled rake, seen at some locks, for dragging weed and other rubbish out of the water.
Legging. The old method of getting boats through tunnels if (as often) there is no towpath. Men lay out on planks to do this and sometimes fell in. At some tunnels men waited at the entrance to be employed for this sole purpose. The coming of engines did away with this strenuous work, but enthusiasts have been known to try it still. They did it often in Dudley Tunnel before it closed for the second time, since no internal combustion engines were allowed there.
Motor. see *Butty.*
Narrow boat. The correct name for the 7ft wide boats able to use the narrow locks of our central canals. The commercial ones were around 70ft long, but many pleasure boats now built to fit these locks are shorter than this. Narrow boats

Kebs by a lock.

were also called "monkey boats" and "long boats".

Roving bridge. Sometimes a special curved bridge was built where the towpath changed sides, to allow the horse to walk over, come back a little, and continue without having the rope unfastened. Also called a turnover bridge.

Shaft. The long pole usually carried on boats for pushing off the side if aground. To be handled with care.

Sill or *cill.* The block of concrete, stone or wood at the very bottom of a lock end, against which the gates stop. You need to be wary of these when "going down" in a lock, so keep away from the gates behind you.

Stoppage. As mentioned earlier, canals have to be closed for certain types of repair, especially at locks. This is usually done in winter, but emergency stoppages may occur in summer. (See Chapter 5).

Summit. Each canal has its highest point, or perhaps more than one high point. This is where water has to be available from reservoirs or pumps, to allow for all the water which passes through locks downwards as boats use them.

Winding. Turning a boat round, especially in a "winding-hole" dug for doing so. Pronounce like the wind that blows. (see Chapter 6).

INDEX

Aire & Calder Navigation 98
Anchor 51
Anderton boat-lift 107
Anglers 49
Anglian Water 158
Animals to see 108
Aqueducts 65
Ashby Canal 102

Balance-beams 70
Barges 20, 86, 98
Basingstoke Canal 158
Baths 26
Batteries 41
Beds and bedding 25, 26, 30
Bicycle 30, 82
Bilges 41
Bingley Five-Rise 89
Bird-watching 108
Birmingham Canal Navigations
(BCN) 107
Blind bends 49, 59
Blisworth Tunnel 64
Boat Show 11
Boatyards 7, 12, 23, 32, 153
Bollards 70, 82
Books 159
Braunston 107
Bridge-keepers 61
Bridge-keepers' houses 93
Bridge-names 59
Bridge-numbers 59
Bridge-slope 59
Bridges - fixed 59
Bridges - moving 47, 61
Bridges - turnover 58
Bridges - variety 91
Bridgewater Canal 158
British Waterways 14, 19, 157
Broad canals 17, 70

Broads 19, 157
Bulbourne 96
Butty 65, 102
Buying a boat 164

Calder & Hebble Navigation 14, 74,
93
Calder & Hebble paddle-gear 74
Caldon Canal 61, 102
Caledonian Canal 67, 85, 89
Calorifier 25
Camping boats 102
Canal "furniture" 14
Canal & Riverboat 10, 158, 164
Canal companies 14, 20
"Canal mania" 14
Canoes 102
Central heating 25
Channel (in canal) 42, 48
Chesterfield Canal 52
Children 47, 61
Clothes 29
Cloughs 72
Clove-hitch 56
Cockpit 27
Commercial boats 19, 86, 98
Commercial waterways 49, 70, 85
Compartment boats 98
Cooker 24
Coventry Canal 96
Cupboards 25

Diesel-powered engine 23
Dredgers 97
Duke's Cut 107

Engine oil 41

Fenders 27, 28, 56, 85
Fishing poles 49
"Flats" (boats) 97
Flowers 108
Footbridges 14
Fossdyke 52, 107
Foxton staircase 89
Fradley Junction 96, 107
Fridge 24, 28
Froghall 102

Galley 24
Gas 28
Gate paddles 72
Gear (put in) 32
Geyser 25
Glass reinforced plastic (GRP) 21,
28

Gloucester & Sharpness Ship Canal
18, 86, 93, 98, 104
Going aground 52
Gongoozlers 10
Goods carrying 12, 20, 98
Great Haywood 107
Great Ouse 51, 70, 88
Ground paddles 72
Guillotine gates 70, 88

Hammer 56
Handrails 27
Hartshill 96
Headlamp 64
Hiring a boat 162
History 12
Holding tank 26
Horn, etc. 49, 59
Horse-drawn boats 102
Hot water 25
Hotel-boats 102

Industrial Revolution 14
Inland Waterways Association 155,
157

Junctions 107

Kennet & Avon Canal 74, 107
Knots 56

Ladders 72
Landranger Ordnance Survey maps
153
Lavatory 26
Leeds & Liverpool Canal 61, 74,
89, 107
Legging 64
Licence 19
Life-jackets 30, 46, 50
Lincoln 107
Llangollen Canal 60, 104
Lock "furniture" 70
Lock shoulders 81
Lock-cabins 94
Lock-flights 67
Lock-footplanks 70
Lock-gates - closing 81
Lock-gates - making 96
Lock-gates - opening 79, 82
Lock-gear 14
Lock-keepers' houses 93
Lock-sill or cill 70, 83
Lock-sizes 19
Lock-staircases 67
Lock-working instruction 32

Locking "downhill" 82
Locking "uphill" 79
Locking - broad locks 85
Lockmaster maps 156
Locks, general see chapters 8 & 9
Locks - anti-vandalism 78
Locks - broad 70
Locks - electric 70, 85, 94
Locks - forward pull 82
Locks - guillotine-gated 88
Locks - narrow 70, 79
Locks - purpose of 67
Locks - staircase 89
Locks - variety 91
Locks on rivers 67

Macclesfield Canal 14, 61
Maintenance boats 97
Maintenance yards 96
Manchester Ship Canal 18, 158
Maps 12, 156
Meeting other boats 48
Middle Level Navigations 67, 158
Mileposts 104
Mirfield 96
Monmouthshire & Brecon Canal 61
Mooring 51, 54
Mooring-pins 32, 56
Museums 14, 64, 104, 158

Narrow boats 20
Narrow boats carrying goods 98
Narrow canals 17, 20, 70
Navvies 14
Nicholson Guides 154
Notices 104

Outboard engine 23
Oxford Canal 61

Paddle-gear 47, 72
Paddle-gear - hydraulic 77
Paddle-gear - working 77
Paddle-gear safety-catch 77
Paddle-rack 77
Peak Forest Canal 61
Pearsons Canal Companions 155
Polythene 52
Pontcysyllte Aqueduct 66
Pounds 67
Propeller 32, 41, 45, 52, 59, 83
Pub-names 92
Pubs 91, 153
Pump-out 26

R Avon (Warks) 107, 158
R Medway 104
R Nene 70, 88
R Severn 51, 94
R Soar 107
R Thames 16, 19, 23, 51, 70, 85, 93, 157
R Trent 16, 51, 52, 86, 94, 97, 98
R Weaver 18, 49, 86, 98
R Wey 158
Rallies 160
Reservoirs 17, 67
Reversing 45, 49, 52, 53, 54, 81
Rivers 13, 16, 18, 51, 85
Rochdale Canal 158
Roof 27, 47, 65
Ropes 32, 54, 56, 81, 82, 85
Rudder 43, 83

Safety 47
Safety - at balance-beam 81
Safety - locking downwards 84
Safety - locking up 82
Safety - when filling lock 82
Safety - with gate-paddles 82
Safety catch 79
Saloon 25
Salterhebble 93
Sanitary stations 26
Sea-going boats 21
Shaft 27, 52
Sheffield & S Yorkshire Navigation 94, 98
Ship canals 18
Ships 98
Shopping 30
Shower 26
Shropshire Union Canal 93, 104
Side-decks 27
Sleeping cabin 26
Slipways 23
Sluices 72
Sound signals 49
Speed-limit 46
Spindles 73
Steel boats 21
Steering 41, 48
Stern gland 41
Stop-gates 106
Stop-planks 106
Stoppages 41
Stopping 45, 54
Stratford Canal 93, 94, 107
Switching off 58
Swivel Effect 43

Table 25
Taps 25
Theatre companies 102
Throttle 32, 41, 45
Throwing a rope 58
Tides 51
Tiller 41
Toilet 26
Tom Puddings 98
Torksey 107
Towing 52
Towpath 56
Traffic lights 85
Trailing a boat 23
Trent & Mersey Canal 14, 64, 74, 107
Trip-boats 102
Tunnels 64, 91
Turning around 53
TV 30

Wardrobes 25, 26
Wash (behind boat) 45
Water authorities 157
Water supply 25, 26, 153
Water-filter 41
Waterman's hitch 58
Waterway guide-books 12, 52, 56, 59, 93, 153-6
Waterway highlights see chapter 11
Waterway Societies and Trusts 157
Waterway system 9
Waterway words 164
Waterways, individual see chapter 12
Waterways World 10, 155, 158, 164
Waterways World Guides 154
Watford staircase 89
Weed-hatch 53
Weirs 51, 67
Wheel 42
Wigan 104
Wind 45, 59, 66
Winding-holes 53
Windlass 73
Windlass - using 79
Wolverhampton 104
Worcester & Birmingham Canal 104

Yorkshire Ouse 51, 52, 98, 158